WIN-WIN ECOLOGY

WIN-WIN ECOLOGY

*How the Earth's Species Can Survive
in the Midst of Human Enterprise*

MICHAEL L. ROSENZWEIG

OXFORD
UNIVERSITY PRESS

2003

OXFORD
UNIVERSITY PRESS

Oxford   New York
Athens   Auckland   Bangkok   Buenos Aires   Cape Town   Chennai
Dar es Salaam   Delhi   Hong Kong   Istanbul   Karachi   Kolkata
Kuala Lumpur   Madrid   Melbourne   Mexico City   Mumbai   Nairobi
São Paulo   Shanghai   Taipei   Tokyo   Toronto

Copyright © 2003 by Michael Rosenzweig

Published by Oxford University Press, Inc.
198 Madison Avenue, New York, New York 10016

www.oup.com

Oxford is a registered trademark of Oxford University Press

Library of Congress Cataloging-in-Publication Data
Rosenzweig, Michael L.
Win-win ecology : how the earth's species can survive in
the midst of human enterprise / Michael L. Rosenzweig
p.   cm.
Includes bibliographical references (p.).
ISBN 0-19-515604-8
1. Nature conservation—Economic aspects.
2. Biological diversity conservation—Economic aspects.
3. Human ecology.
I. Title.
QH75 .R69 2003
333.95'16—dc 21   2002029281

1 3 5 7 9 8 6 4 2
Printed in the United States of America
on recycled, acid-free paper.

Dedicated to Gordon Orians
the very model of a compleat ecologist

Contents

Preamble

There is still time. There is good reason to believe that civilization need not destroy most of the Earth's nonhuman species. The trick is to learn how to share our spaces with other species. If we do so, we won't find ourselves bereft of our plant and animal cousins and hoping for a visit from extraterrestrials to keep us company.

Sharing our habitats deliberately with other species. I call this "reconciliation ecology." The evidence cries out for us to do a lot more of it, and that doing a lot more of it can save most of the world's species. This book will explore that evidence.

The book will also describe many examples of reconciliation ecology, stories of people who have designed habitats for themselves or for their enterprises, and then find out that wild things also use these habitats successfully. Sometimes the sharing is accidental, sometimes quite purposeful. But sharing works. And it is very cheap.

Despite its title, the book may displease some of those who are devoted to "green" causes. They may not trust my claim that we need to end the battle between ecology and economics. But this is a book of science, not theology and not politics. And the claim comes straight from the ecological science of diversity. The science is very clear, and those who care about wild species can do them no better favor than to be guided by it.

Nevertheless, this book is not a signal for environmentalists to surrender their cause to those human beings whose job it is to exploit the Earth. I want our developers, fishers, farmers, ranchers, and tree growers to realize that I am not only calling for environmental peace and cooperation, but also for a radical change in the way they treat the land and waters of this planet. I am not asking them to stop earning a living or making a profit. People and their enterprises will not be denied, and *need* not be denied. But we can avoid a mass extinction of Earth's species without ourselves committing mass suicide.

WIN-WIN ECOLOGY

CHAPTER 1

Reconciliation Ecology

The wolf shall dwell with the lamb,
The leopard lie down with the kid[1]

Today's dominant strategy of conservation biology is reservation ecology: save the Earth's natural habitats. However, in many environments, we have already saved about as much natural habitat as we can. A secondary conservation strategy, restoration ecology, supplements reservation ecology. Restoration ecology tries to return some developed places to a more natural status. But the truth is that even less land remains available for restoration than for reservation. The shortage of land turns out to be a critical problem. Because of it, most species, even those apparently now succeeding in our reserves, will eventually vanish. So we cannot rely on the current balance of conservation efforts. Conservation biology must develop a new strategy if it is going to extend and preserve its successes.

This strategy already exists. I call it *reconciliation ecology*. Reconciliation ecology seeks environmentally sound ways for us to continue to use the land for our own benefit. It follows the words of the Chinese sage who long ago said, "The careful foot can walk anywhere."

Certainly we must not abandon reservation or restoration ecology. We must continue to protect what we have saved. But increasingly, we should turn to reconciliation ecology because avoiding the impending mass extinction will require employing it extensively.

In addition to its primary value as a conservation tool, reconciliation ecology offers a valuable social by-product: It promises to reduce the endless bickering and legal wrangling that characterize environmental issues today. We are all human beings. We share a stake in the world we are building. No one wants it to be sterile and lonely. And no one wants us to destroy our technology and reduce our future to the harsh, subsistence-level lives led by our Stone-Age forbears. Reconciliation ecology gives us a conservation strategy that recognizes these simple truths and unites us in our common goals.

The following example of reconciliation ecology in miniature will help give you an idea of what it means.

The Red Sea Star Restaurant

The city of Eilat, Israel, sits at the tip of the northeastern arm of the Red Sea. Eilat and its neighbor, Aqaba, Jordan, form the extreme western end of a great and glorious biogeographical assemblage, the Indo-Pacific coral reef biome. Its variety of species, its splendor of hues and shapes beggars the imagination. If you have snorkeled or scuba dived there, you know what I mean. If you have snorkeled or scuba dived only in the Caribbean, multiply what you saw there by ten.

As recently as 1960, both Eilat and Aqaba were little more than small village outposts. The coral reef flourished in the clear tropical waters along their waterfronts. But things soon deteriorated. Israel decided to develop Eilat as a deep water port because it still lacked access to the Suez Canal. Construction of port facilities tore into the reef and shipping polluted the waters.

Today, because Israel now has access to the Suez, the ship traffic has dwindled. But a giant tourism industry has taken shipping's place as a threat to the reef. So, Israel has set aside a small fraction of its part of the reef as a national park: Eilat Coral Reef Nature Reserve.

To get to most coral reefs, one has to boat or swim across a lagoon. But the reef of this park lies right at the water's edge, easily accessible to everyone. A luxurious resort hotel stands at one end of the park and from the city center, a municipal bus route will take you cheaply and conveniently to the entrance gate. From there, it is only a 30-second walk to the reef—if you go slowly. Hordes of people come to sun themselves on its beach and swim among its wonders.

The Israel Nature Reserves Authority has met the challenge of such potentially destructive tourism. It carefully supervises tourists at Eilat Coral Reef Nature Reserve and prohibits them from trampling about on the reef—especially in the shallows where the young of so many species grow sheltered from some of their most dangerous predators. I do wish it could also prohibit the loudspeakers of the glass-bottom tourist boats that ply the waters only a few meters from the swimmers, and the blaring of the hotel's musical entertainment system, piped to the beach as an imagined courtesy to its guests. But these detract only from my aesthetic pleasure and only when my head is not underwater. They seem not to diminish the success of the wildlife at all. Eilat Coral Reef Nature Reserve is a joy.

Nevertheless, most of the reef that once lay in Israeli waters is gone, sacrificed to industry and tourism and inadequate sewage treatment. Imagine my surprise then when the travel section of our local newspaper told about the new Red Sea Star Restaurant in Eilat. It would soon open, underwater and surrounded by coral reef! What could those Israelis be thinking of? The only reef left was in the reserve. Were they giving it up?

I should have known better. On a research trip to Israel, my wife, Carole, and I decided to have lunch at this improbable place. We went back and forth along the north end of the reserve looking for a sign. Nothing. Then we found its ad in a local tourist magazine. The article in Tucson's morning newspaper had not been fantasy after all.

So, the Red Sea Star Restaurant existed, but we still could not find it. The road along the shoreline was an unremitting jumble of undistinguished, fairly ugly business architecture. We asked for help and were directed to the worst of it, the corner of the waterfront where center-city commerce turns sharply into a great municipal scar of tourist hotels and shops. Incredulous, we made our way across a short, narrow footbridge to a platform a few meters from shore. On it, barely above sea level, stood a rather small, nondescript structure. We entered.

"Yes, this is the place," we were told. "Just take the elevator down two floors."

Down two floors? I knew how Alice must have felt.

What we saw when the elevator opened made us tingle with awe and disbelief. It could not have impressed us more to have been beamed aboard the bridge of the Starship Enterprise. Half the floor was a cocktail bar, and

The Red Sea Star Restaurant: view of the interior. Courtesy The Red Sea Star Ltd. – Eilat (972-8-6347777).

half, an elegant, white-tablecloth restaurant. Natural light streamed in from the sea—through portholes in the ceiling and fancifully shaped windows that lined the entire structure. Outside those windows was a coral reef full of the gorgeous fish and other lovely animal species that inhabit the reserve several miles away. We had entered an underwater terrarium. We were on exhibit for the fish!

Lunch was delicious. But, as fine as the cuisine was, it could not begin to compete with the space in which we dined. I actually saw a few species that I had never seen in the reserve—despite many visits over the years. In fact, I was so dazzled that I did not even ask the obvious question: Where had this section of reef come from? It had been destroyed decades before. What miracle had resurrected it?

Carole approached the problem with less emotion. She noticed a metal mesh that underlay the coral growth. Was the whole thing bogus? Were branches and bits of coral strung up on a matrix and replaced as they died? Were the fish lured in with supplements of food? Was the enterprise just like the potted houseplants in great tubs scattered about the semi-

Coral reef as seen through a window of the Red Sea Star Restaurant. Courtesy The Red Sea Star Ltd. – Eilat (972-8-6347777).

darkness of a boomer restaurant and rotated out to drink the light on the day before they would have etiolated and sickened? We had to find out.

Reuven Yosef is a pioneer of reconciliation ecology. He lives in Eilat and we were going to see him that evening. He told us the whole story.

Eilat has a fine commercial aquarium and underwater observatory called Coral World. It exists for tourists who cannot snorkel or scuba. Researchers at Coral World had been working to solve an important problem of any aquarium that wants to keep coral alive in tanks of sea water: All species of coral are colonies of animals, and any broken piece of coral soon becomes infected and dies. But the researchers had learned how to treat the coral with antibiotics. After several months of such treatment, a coral fragment would heal completely and could be safely relocated.

Yosef took me to visit the coral hospital behind the scenes of Coral World. There, technicians were treating about forty species of all sorts of coral—branching corals, brain corals, fan corals. Step-like supports, resembling a miniature version of the seats at a sports stadium, sat in large tanks of fresh sea water. On them rested rows of coral fragments,

A ward in the coral hospital at Eilat.
© 2002, Evolutionary Ecology Ltd.

each carefully labeled with its date of "admission." Luckily, one treatment fits all sufferers.

Of course, it is illegal to break off coral fragments from the reserve. Yet, accidents do happen. When they do, the fragments are carefully collected and brought to the hospital. Some, after recuperation, go to the Red Sea Star Restaurant. There, divers wire them to a meshwork of iron cloth where they start to grow, soon covering and enveloping their artificial iron matrix. The fish simply volunteer. As they say, if you build it, they will come.

Presto, reconciliation ecology. A restaurant, designed and built to sit in a novel habitat put together by human beings. Today, its corals come from the casualties of a nearby reserve. Tomorrow, pieces of its coral may serve to repopulate the reserve in the event of some unforeseen catastrophe.

When I first began to notice such conservation efforts, they seemed curious, even odd. Didn't their designers realize that conservation has a duty to focus on pristine environments—or at least on those we are trying to return to some semblance of natural status? But, as you will see, I began to encounter more and more examples like the Red Sea Star Restaurant. Most of these operated on a much larger scale, too. Eventually, I began to appreciate that without reconciliation ecology most diversity is doomed.

The term reconciliation ecology captures the essence of a new outlook for conservationists. Of course, its primary job is to change the way people think about conservation. It does this job by declaring the need to reconcile human uses of our planet with those of other species. But "reconciliation," the word, also accomplishes two other things.

First, it displays its heritage with pride. Reconciliation sounds very much like it belongs in the family of its predecessors, reservation and restoration. In fact, it does. It has not come to unseat them but to join them.

Second, reconciliation also embodies a delicious, intentional, and useful ambiguity. It has both a political and a biological meaning. Although it has no intention of replacing either reservation or restoration ecology, it certainly does mean to replace political discord and enmity with political harmony. Conservation is already difficult enough. Friction will only reduce its efficiency.

But what exactly is reconciliation ecology? **It is the science of inventing, establishing, and maintaining new habitats to conserve species diversity in places where people live, work, or play.** I am not suggesting inventing new habitats in reserves, or in acreage where restoration is going on. I am saying that people now use most of the world's land surface, and we can use it better. We can use it in a way that reconciles our needs with those of wild, native species. Reconciliation ecology is the third 'R' of conservation biology.

To practice reconciliation ecology, we must pay close attention our treatment of the land. We must back off a bit, not on the amount of land we take for ourselves, but on how we transform it for our use. Right now, our footprint is too big. Going barefoot is not the answer, but the time has come to trade in our jackboots for the grace and elegance of ballet slippers. The careful foot can walk anywhere.

We can learn how to reconcile our own use of the land with that of many other species. Maybe even most of them. If they have access to our farm fields, our forests, our city parks, schoolyards, military bases, timberlands, yes, even to our backyards, then they have a chance. If they live where we do, then they have what we have. We shall thus be able to minimize their risk of extinction.

But access to our land is not enough. To practice reconciliation ecology successfully, we must learn what species need in order to get along with us, and we must do that job for thousands of separate species. Then, we must diversify the habitats of our surroundings instead of creating, as we now do, the very limited number of habitat architectures that we have come to like. Every front lawn need not look like a golf course. Every city park need not look like a savannah. Every schoolyard need not look like a desert.

The habitats we create around us will be novel, so the species that we hope they save will not be particularly well adapted to them. Those species evolved in an obsolete world, a world that will not return, a world of coast-to-coast wilderness that has been beaten back to the boundaries of our reserves. Yet, although such habitats may be scarce today, the needs of the species that grew up in them—that adapted to them—remain the same. That is why most native species are best adapted to the scarce

7

habitats of our reserves. To design effective new habitats, we must carefully study the old ones to find out what makes the world of the reserves so suitable. Then we can figure out what is essential and what species can do without. Finally, we can reassemble the critical components into new habitats and landscapes of which we also are a part.

There is a huge difference between what I am advocating and the usual attempt to attract birds to your garden. I am talking about creating self-supporting populations of species on our land. It will take a lot of work. But imagine the result: a vast area of diverse anthropogenic habitats that meet nature halfway instead of trampling her underfoot. Although these habitats would not be ideally suitable for wild things, they would provide enough support to allow them to adapt to us. They would give natural selection the time and space in which to work, and thus could save the overwhelming majority of today's species.

To practice reconciliation ecology we will also need to pool our resources. For a butterfly species, this might mean our banding together in neighborhood groups. No single land parcel may be large enough to contain a self-sustaining population, but 20 might do the trick. Neighbors would join together to select a species and protect it.

For other species, we will need to band together—through our governments—at local or national levels. For migratory species, international cooperation would come into play. I will discuss examples of some of these scales of reconciliation ecology in chapters 2 through 7.

The tasks that we face are all difficult enough, but they pale in comparison to the emotional adjustment we must make. We must give up romantic notions about reserves as wilderness. Yes, we must at last admit the truth: Even in reserves, people can and should be actively involved. That means active management. To make room in our reserves for the species that will need them, we may even have to discourage their use by other species that do not need them, species that we *can* help amid our habitations and our enterprises. It will almost be like admitting that wilderness itself is no more. Yet we must grit our teeth and do it. To do otherwise is to doom most of the very things we want to save.

We could hardly improve on the advice of Dean Acheson:

> It seems to me the path of hope is toward the concrete, the manageable. . . .
> But it is a long and tough job, and one for which we as a people are not particularly suited. We believe that any problem can be solved with a little ingenuity and without inconvenience to the folks at large. . . .

And our name for problems is significant. We call them headaches. You take a powder and they are gone. These pains . . . are not like that. They will stay with us until death. We have got to understand that all our lives . . . the uncertainty, the need for alertness, for effort, for discipline will be upon us.

This is new to us. It will be hard.[2]

Nevertheless it can work, whereas today's dominant conservation strategies cannot. They divide the land into shares, so much for nature and so much for people. This inevitably leads to conflict. And since people are doing the dividing, you can be pretty sure which side will win. We reveal our intentions by calling the shares for nature, "set-asides"—as if we were merely holding them in reserve until we needed them later.

Diversity's best hope comes from what E. O. Wilson calls "biophilia." We all love nature. No one wishes its demise. We merely want to find a way to have our cake and eat it, too.

Lucky us. We can. Reconciliation ecology makes this possible. That Iowa farm family is not about to give up its acreage to be turned back into tall grass prairie anyway. "No worries," says reconciliation ecology. Georgia-Pacific and Weyerhauser corporations are not about to sell their tree farms for the restoration of primary forest, and reconciliation ecology says they need not. Philadelphia is not about to abandon and evacuate itself in favor of eastern white pine, and reconciliation ecology finds ways to make it a "greene countrie towne," as William Penn himself intended when he founded it in 1682. Reconciliation ecology can save species without displacing people or their economic activities. In the process, it can reduce political conflict to a minimum.

Mother Earth, after all, is not really smaller than she was five centuries ago. The Earth abides; the land itself endures. We have just transformed it so much that most species have not yet evolved a use for it.

"Conservation philosophy, science, and practice must be framed against the reality of human-dominated ecosystems, rather than the separation of humanity and nature underlying the modern conservation movement."[3] Today's conservation treats the land as if deriving benefit from it amounts to what mathematicians call a zero-sum game. A zero-sum game is like poker—whatever I win, you lose! If land use were truly a zero-sum game then the only way we could use land for ourselves would be to deny it to other species. But if we can reconcile our use of the land with that of other species, then the land will still be there for both sides.

Reconciliation ecology transforms the zero-sum game of competition into a game that humans and nature alike can win.

Is this too much to accomplish even for almighty man? And what about the costs of discovering how to build reconciled habitats, of actually building them, and of maintaining them in perpetuity? Is it all a mere pipedream? Or was Lewis Mumford right when he advised that, "only the dreamers will turn out to be practical."[4]

Well, if it is a dream, we are already living it. The Red Sea Star Restaurant is but one example of dozens, perhaps thousands of real cases of reconciliation ecology. In this book, you will read about many others. They vary greatly. Some show profits from reconciliation, and some call for better methods to determine its net cost. Some have come from the practice of old, traditional methods of resource use, some from the most modern concepts of design and technology. And these reconciliation projects have been accomplished in rich, poor, and in-between countries, in all sorts of habitats, and at levels from entire national governments, to large private landowners, to individual homeowners. The truth will make you an optimist even if you continue to be the staunchest, most unwavering pragmatist in the world.

CHAPTER 2

Landscape Architecture for the Third Millennium

> Looking for fish? Don't climb a tree.
>
> Old Chinese Proverb

> Habitat.
> If a creature gets into the right place,
> everything else is likely to be easier.
>
> Heerwagen & Orians[1]

The pocket mice did it.

Pocket mice are wild rodents of North American deserts and semi-arid landscapes. Seeds form a very important part of their diets. These mice look for seeds on the desert floor and then cram their winnings into fur-lined pockets inside their cheeks. Their name comes from these pockets.

After accumulating a good-sized stash, they return home to their burrows and deposit the seeds into a storage chamber. Some also dig special pits all over the desert floor for storing the seeds.

Tiny little guys even as adults, they weigh from a quarter of an ounce (eight grams) to just over an ounce (a bit more than 30 grams), depending on the species. They are more closely related to squirrels than to house mice, and I think they have a lot of personality.

In the 1970s, I was working in Arizona to discover how so many species of pocket mice—there are dozens—could stay in business. Just like

Desert pocket mice (*Perognathus penicillatus*). This is the very species whose sophisticated habitat preferences I describe in the text. From a painting by Allan Brooks.

human businesses, you see, a natural species must have an angle, something that it does different from others, something to protect it from the often harsh competition of similar species.

After only a few weeks of field study, I began to notice that many pocket mice lived in habitats that were consistently different from those of others. In particular, some species seemed to associate with patches of grass, while some preferred to forage under the cover of small desert bushes.

I decided to run an experiment to see whether these habitat differences really mattered by going into the space of some pocket mice and re-landscaping it. Then I'd see if the mice stuck to their homesites or moved away.

This experiment had two goals. First, I wanted to see if I'd been reading the mice correctly. I mean, think about all the ways there are to measure a habitat and change it. Are bushes and grass really important? As I mentioned, all pocket mice live in burrows that they dig in the desert floor. Maybe the bushes grow only on a certain kind of soil, and the mice don't care a bit about the bushes, but really need the soil in which those bushes grow.

The second experimental goal related directly to reconciliation ecology. I wanted to find out if the mice actually choose their special habitats. Unfortunately, I don't speak any dialect of pocket mouse, so I couldn't exactly ask them. But I could study the rate at which they disappeared after I cleared away some of their brush. If they vanished slowly, I'd know they were getting picked off by predators or suffering from an inadequate food supply. But if they vanished all of a sudden, it would mean they were actually pulling up stakes and choosing to live elsewhere.

During the first summer of these bush-clearing experiments, the pocket mice ignored our landscape changes entirely. But instead of concluding that bushes do not matter, we decided to try again the next summer with a slightly different landscaping plan.[2]

We guessed that if bushes mattered, it was because they afford some shelter to which an exposed pocket mouse could retreat whenever it sensed danger. Maybe we had not cleared away enough bushes in the first summer. Maybe the open spaces we created were not big enough. In the first summer, a pocket mouse always had a bush within 26 feet (eight meters). Maybe that was close enough to satisfy its need for safety. So, this time we cleared away more bushes, doubling (to 52 feet) the distance between foraging sites and cover.

I cannot say how fast those pocket mice decided not to keep using those re-re-landscaped foraging sites. Too fast for me to measure. When we finished our work, I suspect there were a lot of pocket mouse conversations that went something like this:

"Hey, Maude, do you see what happened to the back 40?"

"Yep, Charlie. Looks like it was ruined by a hurricane. Anyhow, we sure can't use it anymore."

And they didn't.

And so, these half-ounce beasties taught me to respect the cognitive power and discriminatory ability of "lower forms" of life. Biologists had long known that all forms of life have special habitat needs. What had not been previously known was the sophistication with which some animals assess their environments and choose the places where they will live and work.

Active habitat selection like this is a two-edged sword for reconciliation ecology. One edge tells us we have some mighty picky customers to satisfy. The other suggests that we can rapidly perform experiments to identify the critical elements of habitat design that will at least get our customers to give reconciled habitats a try. Think of it as market research for wildlife.

Human Habitat Selection

Sometimes we humans forget that we have an animal side to our nature. It is a side particularly easy to deny when it comes to habitat selection. Unlike animals, we seem able to live anywhere. Hot or cold. Wet or dry. Forests, plains, deserts, and seashores. It is very difficult to imagine a human family leaving a place because their shrubs got uprooted by a windstorm!

But look more closely at where we live. In truth, we do not actually adapt to all those habitats. Instead, we modify them to suit our own desires as much as we can. We try to control their temperatures and humidities within narrow bounds. To extend the daytime, we illuminate them with electric lights. And we landscape the devil out of them.

If you look at landscaped human habitats around the world, you may be struck by how much they resemble each other. Some expanse of grass, a scattering of small trees—often with multiple trunks—and some amount of surface water. W. Whyte[3] reported what people want in their parks— trees and grass. Ask a realtor which she would rather sell: a splendid large house on scraped-bare land capped with asphalt, or a more modest house surrounded by lawn, a few trees, and next to a brook? Ask the developers and architects who use drawings and scale models to sell their projects why they include such elements. If they deviate from the formula, it is to substitute a lakeside for the brook. Ask China why its Sung dynasty was so proud that in its capital city, no street lacked the sound of water and the scent of flowers.[4] And ask Europe why it peppers its charming, older, denser, hard-core urban centers with so many extravagant fountains?

Facing page, top Sibelius Park, downtown Helsinki, Finland. Trees and grass. If you could turn around and look behind you, you would see a magnificent fountain. © 2002, Evolutionary Ecology Ltd.

Facing page, bottom Fort Lowell Park, Tucson, Arizona. Trees and grass and a pond. © 2002, Evolutionary Ecology Ltd.

Sonoran Desert natural landscape just outside Tucson, Arizona—brimming with wild things, but it's no place to mow grass. © 2002, Evolutionary Ecology Ltd.

To ecologist Gordon Orians, it sure looked like we adjust our habitats to satisfy our fundamental, deep-rooted animal nature. So he and a psychologist, Judith Heerwagen, began a systematic study of human habitat preferences. What they found speaks to a profound unity among us. We seem to have grown up as a species that did best under well-defined circumstances and we have never outgrown our desire to live in them.

Orians and Heerwagen[5] uncovered the evidence of our preferences in several ways. They researched old drawings of landscape architects and artists. They studied how people decorate their offices, comparing the choices of those office workers without windows to those with windows. (The latter hung the abstract pictures; the former, more naturalistic scenes.) They compared natural forms of trees with those bred for gardens. And they showed a lot of people a lot of pictures and asked them to rank their attractiveness. What they discovered rings true.

First in our hearts comes a grassy savannah habitat. A few trees growing in a rather open vista satisfies us at some deeply subconscious level. You could say we just love wide open skies. And the trees? We prefer our trees with several trunks, or at least with only a short trunk topped by a wide branching structure (perhaps to give us shelter from sun or rain). Orians guesses that multi-trunked trees may also be easiest to climb should we be surprised by the odd lion prowling the grassland for a tasty 100-pound morsel. Maybe so, but in any case, we like such trees.

They also discovered that we exhibit slightly different habitat preferences depending upon our sex. Men prefer the more open vistas of the savannah. Women like a more sheltered, lightly wooded corner. Orians and Heerwagen had actually predicted these differences based on the likelihood that primitive people were hunter-gatherers, with men emphasizing the hunting and women the gathering.

Of course, we like water, too. So we want our savannah on the edge of a lake or a river. And we want it full of grazing animals like deer or sheep. After all, such habitats did once supply us with both food and drinking water.

And we love flowers. In fact, we spend millions on them. We offer them as tributes. We give them to each other in times of greatest crisis or joy. Our passion for them may reflect a need for the fruits and other foods that they signal. Or, you may believe that our passion for flowers reflects God's wisdom and benevolence in creating us to appreciate diversity.

What is the significance of human habitat preference? Just this: Very few species like what we like. In fact, very few can even survive in the habitats we like. That is why so few species now live with us. If people continue to have their unbridled natural way, survival will only get tougher for almost everything else.

And people can be quite pushy about their preferences. Consider the covenants and restrictions attached to many people's homes. Keep your grass mowed. Keep weeds out. A friend of ours once belonged to her neighborhood's volunteer inspector corps. She took her job of ferreting out untidiness very seriously. After all, an untidy house is worth less and lowers property values in the neighborhood.

Helmut Zwolfer once told me why he moved back to Germany from Switzerland. He studies and appreciates certain fruit flies that require specific weeds to eat. So, he encouraged those weeds in the garden of his Swiss home. Heaven forbid! His neighbors would not let him be. They did not care for his habitat and made sure he knew about their displeasure. Constantly. He returned to Germany where people are somewhat more tolerant.

The U.S. National Wildlife Federation must have encountered this problem, too. For almost three decades they have sponsored a campaign called Backyard Wildlife Habitat. It encourages people to bring nature to their own home. Consider some advice from the book, *Gardening for Wildlife.*[6]

Gardening for Wildlife teaches people how to create backyard habitats. On page 61, it has a section about the steep lawn, which is both difficult and even treacherous to mow. The authors encourage people to turn such a lawn into a patch of unmown prairie, brimming with wildflowers and native grasses. Next comes an implicit warning: "Put up an 'American Prairie Garden' sign in front of your new landscape, explaining the plants and wildlife in your prairie. Your neighbors will be more tolerant of your new garden if they know what's going on."

What a great tip! If you let weeds grow in your suburban backyard, you had better believe your neighbors will soon be complaining about you. And only the most courageous will complain to your face.

Herbert Bormann, Diana Balmori, and Gordon Geballe wrote a revolutionary little book called *Redesigning the American Lawn.*[7] Among its jewels are four stories of people who tried to foster something different from the usual sward of pseudo-savannah. Three of the four report some friction between them and their good neighbors.

BACKYARD WILDLIFE HABITAT®

HABITAT®

NATIONAL WILDLIFE FEDERATION®

In one case, the perpetrator of the non-lawn happened to be a professional botanist. He soothed his neighbors with an informal and friendly education, delighting them with descriptions of the natural history of the wildflowers that were invading his land. Soon, some even joined him in his heresy.

In a second case, the "offenders" failed to clip their "lawn" down to 12 inches—the height required by county ordinance. Under neighborly pressure, the county issued a citation, but in the end, the county had to change its rules. The wild meadow stayed.

The third case has real charm perhaps because it reflects the deep-rooted mannerliness of southeastern American culture. An acre of disorder-by-design surrounded the home of an insect-loving biologist in Athens, Georgia. Beset by complaints, creative Athens officials declared it a bird sanctuary, and put up a large sign so informing the public! End of problem.

How tidy we are. As long as we know it's a bird sanctuary and not an unkempt lawn, it is acceptable. Maybe it even increased local property values. When Shakespeare penned, "What's in a name? A rose by any other name would smell as sweet," his was just a cry of exasperation. He knew very well how important labels are to people.

It is pointless to bristle. It makes no difference that this is the land of the free. And no one cares that they have not been asked for an opinion. *Homo sapiens* is a pushy, nosy species. And it knows what it likes in the landscapes around it. Reconciliation ecology will have to adjust to each species it works with. And that includes our own. We shall have to work very hard to attain the golden day when Bormann, Balmori, and Geballe get their wish[8] and "Dandelions and crabgrass . . . become things of beauty and admiration and brown spots (are) evidence of natural cycles."

Up Close and Personal: Backyard Habitats

The National Wildlife Federation's Backyard Wildlife Habitat campaign provides a beacon for those of us who believe that reconciliation ecology can pervade our homesites.[9] In its nearly 30-year existence, it has enrolled more than 20,000 private little patches of nature. They vary in area from a few acres to much less than an acre. All have as their goal the creation of a modified human habitat that speaks to the needs of at least some local wildlife.

Backyard habitats form woods and prairies, streams and deserts. They attract birds, mammals, frogs, and butterflies. They support wildflowers, trees, mosses, and grasses. And they discourage the introduced pests that so often interfere with natives.

Were you to examine the manual on backyard habitats,[6] it would impress you in many ways. First, it is full of pictures of the glories it can lead to. Second, it is practical; it has directions, timetables, and plans. There is even an application form! Third, it gives a brief introduction to the biology of the natural habitats that will be emulated. Yet, despite its zeal, it is not in the least theological—for example, as a means to reduce or eliminate mowing, it often recommends intense mowing for a limited time. Nature does its business in funny ways and this book knows it.

Wisely, the backyard wildlife habitat strategy does not ask us to abandon our properties to wild things. In fact, if you scan its landscape designs, you will see that each one pays attention to the needs and desires of people. It harmonizes our image of a perfect habitat with those of many other species. Sometimes it accomplishes this by setting aside small patches of lawn overshadowed by shade trees and harboring a comfortable bench or two in the midst of the lawn. Or sometimes it calls for a pleasant pathway to thread through a patch of woodland. Again, we detect in the backyard strategy both the absence of dogmatism and the desire for reconciliation. Backyard habitats may be patches of nature, but they are also homes for their owners.

Diversity would stand a much better chance if we did no more than follow the backyard habitat idea. But, good as it is, we can make that idea even better. Here's how:

- Backyard habitats are generalized—woods, meadow, desert scrub, and so on. But, to be its most efficient, reconciliation ecology needs to set up specific habitats for well-identified species. Otherwise it would be guilty of ignoring the rich and subtle variety of habitats that allow so many species to flourish.

- Backyard habitats often hope to attract wildlife. To do so, a habitat may be located across from a large park or wildlife preserve. In contrast, reconciliation ecology wants habitats that can actually support wild species on their own. That may seem a minor difference at first, but only by supporting populations over the long term can we decrease extinction rates.

21

Both those additions to the backyard concept present difficulties. First, how do we design the specific habitats? Second, how do we use the postage stamps that are our house lots to achieve self-sustaining populations of wild species?

The answers will come from research, coordination, and association. We ecologists have shown repeatedly that we know how to figure out what a species likes and needs. The pocket mouse example at the beginning of this chapter is only one of many cases. But we must also involve the landscape architects. With their fund of skills joined to ours, we can all learn what must be done.

Finally, people must join together. I envision neighborhood societies for the support of particular species. That butterfly, which is merely attracted to a patch of *Caesalpina*, may be supported perpetually by 1,000 such patches not too far from each other.

City Parks and Gardens

André Le Nôtre conceived and created the gardens of the Versailles palace under commission from King Louis XIV. They took 22 years to build and disappoint no one. The gardens of Versailles constitute perhaps the world's most extravagant example of a quasi-natural, human-designed habitat. Sweeping vistas. Geometrical displays of flowers and hedges all carefully weeded and scrupulously pruned into shape. Versailles is landscape architecture on such a grand scale that we call it "art."

But Versailles makes no claim to supporting many wild species. And it does not. It is a fantasy habitat, probably the subconscious formal expression of all the natural habitat features we humans want. It is a kind of caricature, an unbridled, infinite-budget exaggeration of the ideal human habitat.

Thousands of miles away lies another great public garden, Golden Gate Park in San Francisco, California. Actually Golden Gate is a 410-hectare system of gardens. (A hectare is about 2.47 acres, so Golden Gate has 1,013 acres.) Some gardens, especially its Japanese Garden, are virtually as formal as Versailles. Others, especially those near its western end, give the appearance of being natural.

But appearances are indeed deceiving. Golden Gate was invented by humans. It began as a system of sand dunes. Humans created its mead-

A view inside the gardens of Versailles. Its extreme angularity and formality tells you all you need to know about the goals of its architect, André Le Nôtre. The green plants were his bricks, the flowers his palette. Nature had nothing to do with it. Paradoxically, it almost seems as if no human deed—not even the total destruction of an area for a shopping mall—could more emphatically express man's domination of and disregard for nature. Photograph by Patrick Dorangeon, École publique Saint Vincent d'Ardentes.

ows, and keep them mowed. Humans dug and lined its ponds, filled them with water, and then introduced about eleven species of water-fowl, mostly exotics. Humans planted its trees and shrubs, many of them also exotics. Golden Gate Park is as much a fantasy as Versailles!

But the spirit of Golden Gate Park is not the spirit of Versailles. Versailles flaunts its unreality proudly; Golden Gate Park conceals it deceptively.

We have good company in being deceived by Golden Gate Park. Many species of birds also see it as set of natural habitats. Joseph Mailliard listed 111 species of birds in the park (not including the introduced waterfowl) when he compiled his handbook of its birds.[10] Most or all of these may still be seen there.

Carla Cicero restudied the park's birdlife[11] in the 1980s. She focused on five ponds in the western half of the park and counted 84 species. But she wanted to know more than the species list. She wanted to know what features of these artificial habitats made them desirable to various species of birds. She asked that question as a landscape architect, fully prepared to recommend habitat changes that would support even higher populations and greater diversities of the park's native species.

And why not? These ponds are important sites for human recreation—even including model boating. But no one will cry interference if we help the song sparrows by allowing a few more shrubs in the understory around the ponds. And no one should mind if more vegetation overhangs the ponds to help the kingfishers and the black phoebes. Changes like that will have the power to make a master illusionist out of Golden Gate Park.

Don't misunderstand me. I don't want to start a witch hunt. Formal gardens are indeed works of art, and they are scarce enough that we can let them be. But we need no more of them. We need to reconcile our parks to their true capacity for supporting wild species. Versailles is extraordinary and should remain so. But let us make our Golden Gate Parks a commonplace.

The Green Roofs of Berlin

Reconciliation ecology in core urban areas? Do not write off the possibility—even in the largest ones. For example, in Berlin, people have planted gardens on many large rooftops. They are not yet reconciliation projects, but some could be. Not the ones planted in soil over a foot deep. They require intense watering of ordinary bushes, flowers, or even vegetables. But roof gardens that have shallower soil and take little maintenance would make ideal targets for reconciliation. None of these gardens is watered, fertilized, or mowed. They come in three soil depths.

Roof gardens with less than two inches of soil have mosses. Those with two to four inches grow fleshy-leaved colorful *Sedum* species. And those with four to six inches support grassy mini-meadows.

Has anyone ever thought before of these roof gardens as defenders of diversity? And why not? Manfred Köhler[12] put out 90 shallow containers on the roof of his building at Berlin's Technical University. Each had an area of only 1.2 square meters (12 square feet) and used but one soil mixture. Yet, of the 22 species that he sowed in his simple garden, he found that seven succeeded. They germinated and flowered and set seed. More-over, 24 other species blew in on the wind and had the same success.

Why not enlist the roofs of the world's cities in the campaign to save species? Who knows how many we could help if we varied the design of the gardens? In addition to growing them in the world's many climates, we could vary their soil and their exposure to the sun. We could even burn some gardens periodically to protect the species adjusted to such disturbances.

People planted the roof gardens of Berlin to enjoy their beauty. They have learned that those gardens actually improve the climate of a city and help it to conserve water by reducing evapotranspiration. What now prevents people from deriving yet another benefit from roof gardens? What keeps us from using them also to preserve some species of plants?

The lesser kestrel, an endangered falcon species, nests and thrives in the roof tiles of Jerusalem. From a painting by Tuvia Kurz. Courtesy of Tuvia Kurz, the International Center for the Study of Bird Migration, Latrun, Israel and the Society for the Protection of Nature in Israel (SPNI).

CHAPTER 3

Prometheus in the Pinelands

Everything that can come through fire,
you must pass through fire.
Then it will be pure.[1]

Before Prometheus gave fire to Homo sapiens, he had already given it to a very long list of other species. You see, fire destroys habitats, true. But it also creates them. A roaring fire transforms the land so completely that the survivors might as well be on another planet. Shade vanishes and light streams in. Nitrogen, phosphorus, and potassium—the basic nutrients on which life depends—flood the landscape in abundance; the fire sets them free from their organic prisons in wood, in stems, and in leaves. Fire yields mass death, and mass death yields opportunity.

For a very long time, we Westerners saw only the deaths. We fought fires wherever they tried to start. We invented Smokey the Bear to help us perpetuate our one-sided view. We enlisted the military to sweep over fires in old bombers, releasing smothering slurries and choking off an important natural force of renewal.

It had not always been so. North America's first citizens, especially those that lived on its prairies, well knew the virtues of fire. They set fires intentionally, keeping the tall grasses thriving for the bison on which their lives depended.

Even more so did Australia's first citizens. For 40,000 years, aboriginal Australians started fires regularly. They kept track of when they had set them. Then, during the full force of the renewal (usually a few years after they had set a fire), they would revisit to hunt and harvest the bounty of the animal life their fire had brought forth. In a very real sense, they were fire farmers, using fire in much the same way we use plow and fertilizer. Of course, no one has ever accused modern Australians of interrupting the aboriginal tradition of fire setting.

Many plant species depend on fire to survive their competition with other plants. The aboveground parts of some of these fire-loving plants may burn completely, but they resprout from underground organs protected from the intense heat. Prairie grasses are an example. Other species, especially certain trees that grow where lightning is common, have thick bark that protects even aboveground parts from all but the hottest of fires. Finally, others, like the shrubs of the fynbos heathlands near Capetown, South Africa, have actually evolved to burn like hell. As if they know that fire is their friend, they produce flammable biochemicals that turn them into torches once a fire starts. They die, but their seeds germinate and take advantage of the flush of nutrients.

Some species of pine tree have the thick bark of fire resisters. Pitch pine, slash pine, jack pine, Apache pine, and Chihuahua pine are among them. One, the southeastern U.S. longleaf pine, once grew straight and tall in massive expanses that beckoned the chainsaws of commercial loggers. Their trunks made ideal ship timbers, not only for their length and straightness, but also for their resistance to rot. At one time, people also drew away their sap for turpentine.

Longleaf pine forests, with their characteristic understory of wiregrass, used to spread over huge fractions of states along the coastal plain from Texas to Virginia, covering some 90 million acres.[2] Most of these forests have now been timbered and turned into forest plantations. To replace the longleaf pine, logging companies often planted slash pine and sand pine because they yield higher profits. The U.S. Department of Agriculture's Forest Service estimated[3] that even these exploited acres had declined 16.3 percent in the 15 years from 1977 to 1992, and that only 153,200 acres of longleaf pine/slash pine forest reserve remained. (Slash pine gets combined with longleaf in timber-oriented inventories.) That amounts to 1.1 percent of the 1992 forest. It is an even smaller percentage—0.2 percent—of what George Washington would have surveyed in the mid-eighteenth century. Eglin Air Force Base estimated[4] that, in 1992, as little as 5,000 acres of old-growth longleaf pine remained from its entire original 90-million-acre range. That is only 0.006 percent of what it once was. Longleaf pine forest is one of the great loser habitats in the United States. It is doing much worse than average. We have stripped it to a shadow.

Without a continuous, extensive forest to spread the fires that lightning sets, today's fires in longleaf pine forests are smaller than those of yesterday. In the past, an average acre burned every two to five years. Today's average acre burns much less often. And so, the longleaf pine is starved for fire. Its competitors, especially hardwoods, sprout up and take over the sandy tracts that rightfully belong to pine. Even where timbering has not cut down the pines, such as in the longleaf reserves, they are disappearing.

And they are not going alone. Longleaf pine defines a whole ecosystem. The sandy soil that supports them also supports dozens of other species, plant and animal. Red-cockaded woodpecker. Gopher tortoise. Pineland wild indigo. Toothed savory. Many of these can survive nowhere else. The pinelands are a veritable Noah's Ark of diversity.

One great stand of longleaf pines remains. Its location: Eglin Air Force Base just east of Pensacola in the Florida panhandle. Eglin encompasses 463,448 acres and most of that used to be pineland. Eglin uses less than 15 percent of its area to do its various military jobs and house some of its people. These functions are scattered about the base in 36 separate clusters. The base develops and tests various weapons for the military, and many of these are meant to explode, so the military policy of fencing off the base seems particularly appropriate. Yet, many thousands of people live on the base in its 2,380 homes, and thousands of others buy permits to camp, fish, and hunt there.

For Eglin Air Force Base, the challenges began toward the end of the Reagan presidency. Eglin was designated by the Air Force to conduct tests of one part of President Reagan's Star Wars initiative, the controversial program to shoot down incoming ballistic missiles. I have no notion what exactly Eglin was meant to test, but the particular weapon or weapon component was apparently not something a bomb squad would be happy to explode in a garbage can. Indeed, its very testing might have produced effects well beyond the test range itself. That possibility captured the attention of the U.S. Fish & Wildlife Service (USFWS).

USFWS has the responsibility to enforce the Endangered Species Act (ESA). Red-cockaded woodpeckers are an endangered species, listed soon after Congress passed the act in 1973. The woodpeckers live on Eglin Air Force Base, and USFWS decided that the tests might jeopardize their survival.

During one crucial meeting, the Department of Defense (DOD) responded with what looks like a power bluff. At a workshop in February 1992, a DOD representative emphasized the primary mission of the nation's military bases, and went on to warn that unless endangered species management programs could be integrated with military operations, "DOD (might) request and receive an exemption from the Endangered Species Act."

Perhaps in 1982 the bluff would have worked. But not in 1992. No, by then, the military was encouraging environmental work on its bases. The tide had truly turned, and the implicit threat of that employee had become a relic of a bygone era. Eglin's own managers at the workshop countered the negativity right away. They saw no reason to assume that the needs of red-cockaded woodpeckers could not be reconciled with those of the military. The military moved the tests somewhere else. Eglin had become environmentally sensitive. But the story is not over.

In 1990, Eglin did a survey of red-cockaded woodpeckers and earmarked money for endangered species management. In 1991, the base did a rare-plant survey, and, in 1992, a survey of rare lizards and amphibians.

In 1992, the Air Force invited the Florida Natural Areas Inventory (FNAI) of the Nature Conservancy to survey all Eglin's rare plants and animals. FNAI found more than 90 species alive on the base. A number of these are of national concern. Several are already listed as rare or threatened under the Endangered Species Act, such as red-cockaded woodpeckers and Bachman's warblers.

A hundred years ago, about 78 percent of Eglin was sand hills covered with longleaf pine forest. Much of the rest also grew longleaf pine. But previous generations—especially during the twentieth century—destroyed almost all of the primordial pineland. Not only did they cut longleaf pine down, they put out wildfires in its midst. The scarcity of fire cancelled longleaf pine's most important trump card, and encouraged massive encroachment by scrub oaks and sand pine. In 1939, sand pine grew on only 7,000 acres of the base, but by 1979 it covered 60,000 acres. Moreover, once they "harvested" the longleaf pine, Eglin's foresters usually replaced it with slash pine or sand pine, because—until 1978—no commercial nursery supplied longleaf pine seedlings—there was just no market for them. From 1950 to 1977, they artificially planted more than 41,000 acres, and only 12 percent of it was longleaf pine.

Furthermore, much of the remaining longleaf pine was in trouble. Its understory—instead of wiregrass—was a tangle of plants that did not belong, protected by insufficient fires during the growing season. Oaks of several species were choking out the longleaf pine seedlings. Only four small areas of old-growth longleaf pine remained. These totaled 1,712 acres and they were heavily infested with various species of oak trees in the understory. The pines were doomed. Something had to be done.

After much consultation and discussion, the Air Force in 1993 put together a five-year plan for restoring the forest as much as possible. They removed large numbers of the sand pines and slash pines. They planted more than three million longleaf seedlings. And they are burning the understory of the forest during the growing season, when it does the most good. In 1996 alone they burned more than 25,000 acres. Eventually, they hope to burn 50,000 acres per year.

It is working. Longleaf pine now dominates more than 200,000 acres of Eglin Air Force Base. Joining Emily Dickinson, we may well and truly say:

No vacillating God
Ignited this Abode
To put it out.[5]

Meanwhile, ecosystem managers have not ignored the plight of the pineland's endangered species. The managers direct their most intense efforts toward the red-cockaded woodpecker itself. This species depends absolutely on mature longleaf pine. Even in the presence of such pines, the woodpecker will actually pull up stakes and abandon a forest that has too much oak growing under its old pines. Picky pecker.

A seven-inch-long black-and-white bird, the red-cockaded woodpecker does not have quite

Burning the understory of a longleaf pine landscape in Eglin Air Force Base, Florida. Photo by Brenda Biondo.

the glitz of its huge, spectacular, and now extinct relative, the ivory-billed woodpecker. Its name suggests a blaze of crimson brilliance, but in fact its red cockade is an obscure little patch of red feathers confined to the temples of the males. Nonetheless, it is warm and feathery and has some interesting habits that endear it to those who know birds.[6]

It lives with its relatives in small clans, each containing only a single mated pair. The others are either this year's crop of babies, or older relatives. The older relatives actually help the parents with their rearing chores—brooding and feeding.

Of all woodpecker species in North America, the red-cockaded woodpecker is alone in excavating its nest holes in live trees. It will not use a dead trunk or an artificial nest box. The red-cockaded woodpecker does accept some help,

Red-cockaded woodpecker by George Miksch Sutton.

however. It often chooses old longleaf pine trees that have been attacked by a fungus.

The red-cockaded woodpecker used to be a common bird, but its devotion to longleaf pine has nearly cost it its existence. It was appointed to the rare and endangered species list in 1973. Nevertheless, its population keeps dropping. During the 1980s, the red-cockaded woodpecker lost another 23 percent of its individuals. Have we lost this battle? Is it time to say, *Adios, amigo*?

Eglin's wildlife managers addressed the problem of the red-cockaded woodpecker primarily by restoring longleaf pineland habitat. But they did not stop with that. They learned how to drill artificial nest cavities in the trunks of the longleaf pine. If they had waited until these trees were old enough to get fungal rot, there might well have been no more red-cockaded woodpeckers left to take advantage of the opportunity.

Evidently, the artificial holes work. The red-cockaded woodpeckers now nest in 30 percent of the holes, and the woodpecker population has begun to grow. From 1995 to 1997 it increased 6 percent. That may not sound like a lot, but it is in the right direction for a change. Approximately 523 red-cockaded woodpeckers now live on Eglin.

32

Drilling woodpecker holes in young longleaf pine trees, Eglin Air Force Base, Florida. Photo by Sean Livesay.

What makes the management activities at Eglin Air Force Base reconciliation ecology? Is it not simply another example of restoration ecology? Not exactly. While the methods of restoration ecology have been crucial at Eglin, the truth is Eglin is reconciliation ecology writ large.

Why? Most important, the Air Force has not returned its property to the country's bank of wildlands. Eglin is still an Air Force base. It flies planes, tests munitions, and performs all the other functions for which it was set aside. But its human uses no longer ignore the wild species that can and do live inside its fences.

In addition, the Air Force intervention has been crucial. Simply fencing off the pinelands merely slowed their demise. Now people themselves had to start the growing season fires that allow the pinelands to flourish. If the managers were to relax their grip, we could expect all the old problems to resurface.

Finally, part of Eglin's strategy is to manage for recreation, including hunting and fishing. No one has proposed keeping human visits to the same low intensity as in a restored reserve. Even logging continues, but without ruining the longleaf pineland. "Goal 11" of the base's 1993 plan deserves full quotation: "Produce products and services in harmony with restoring and maintaining the long-term sustainability, diversity and productivity of the ecosystem." Notice the words "in harmony." They boldly proclaim the intent to reconcile. What could be clearer? Eglin forest managers intend to be forever drilling holes for red-cockaded woodpeckers in the longleaf pine stands they harvest for timber.

On the other hand, the environment that Eglin wants to re-create in the old-growth pineland is not novel. Thus, that part of their plan does amount to restoration rather than reconciliation ecology. Old-growth pineland may be archaic, it may be doomed without management, but it is not new. Therefore, the part of the plan that focuses on old-growth pineland keeps the efforts of Eglin Air Force Base from being pure reconciliation ecology.

So what. Do you care? I don't. What interests me is conservation of diversity. If the methods and philosophies of reconciliation and restoration cannot be altogether segregated, so be it. Let them cooperate for success.

The Air Force's success at Eglin inspired a cooperative 1996 agreement with its neighbors. Called the Gulf Coastal Plain Ecosystem Partnership, it involves Eglin and six other large landholders in Florida and Alabama with property that adjoins Eglin. Together they control some 840,000 acres of longleaf pine habitat. One is the Nature Conservancy itself. One is a private corporation. And the rest are federal, state and local government agencies primarily concerned with resource management. No doubt they will seek to apply methods similar to those that Eglin has pioneered. I trust they will also succeed.

The Nature Conservancy is not resting on its laurels at Eglin. Its Florida Natural Areas Inventory has now surveyed Camp Blanding Training Site, a 73,000-acre, Florida Department of Military Affairs and Army National Guard facility. In the pinelands, the inventory workers found 18 species of rare vertebrates, plus several rare invertebrates and plants. The pines are in deplorable shape. The inventory recommends the same sort of intervention as succeeded at Eglin.

Eglin may be spectacular, but it has good company. The Department of Defense controls about 25 million acres in the United States; much of it is available for reconciliation. Biondo quotes Doug Ripley, a retired Air Force officer, to explain why much of the military's land is in such good shape compared to other property: "We didn't muck it up; we put a fence around it and we kept people out. And that ended up being a pretty neat thing for the environment."[7]

Accordingly, the Nature Conservancy has a cooperative agreement with the Department of Defense. Under it, more than 200 conservation projects are taking place on more than 170 bases in 41 states. These include the Army's Fort Hood (Texas), the Navy's Boardman Naval Weapons Training Center (Oregon) and the Marines' Camp Pendleton (California).

Even NASA (National Aeronautics and Space Administration) is getting into the act. It radio-tagged gopher tortoises at Kennedy Space Center, Florida. Yes, a natural population of this troubled, sand-hill-loving, longleaf-loving turtle actually lives on the space center's grounds amid the rockets and the high technology. NASA is trying to discover how to improve their condition.

The plight of the longleaf pine forest has attracted a second pioneering program. It stems from some discouraging facts connected to the Endangered Species Act.

- A large number of the country's threatened and endangered species live mostly or entirely on private property.

- Threatened and endangered (T&E) species do much worse if they live entirely on private land than if they live entirely on federal land. Of those T&E species that live entirely on private land, only 3 percent are improving, whereas 18 percent of those that live entirely on federal land are improving. The disparity is little better if we focus on species that have stable populations: 19 percent live entirely on private land; 39 percent live entirely on federal land.

- Private landowners do not like to permit censuses of T&E species on their land. That is one reason we know so little about the status of 51 percent of T&E species that live entirely on private land.

- The ESA does not require or encourage landowners to do anything to prevent the population of an endangered species from falling. If landowners decide on their own to do something beneficial for a T&E species, they must pay all the expenses themselves.

- If landowners improve the quality of their property for rare species, and succeed in attracting a T&E species to it, then the ESA places new restrictions on their authority to use their land.

- If property supports a species that landowners guess will one day get "listed," the ESA places no restrictions on what the landowners may do. They may, for example, destroy all natural habitats to avoid ever harboring a newly declared T&E species.[8]

Can you believe it? Despite the importance of private lands, the ESA's relationship to the private landowner is altogether pernicious. Whoever ruins the land for wild creatures goes unscathed and continues to have the unrestricted right to exploit it. Meanwhile, whoever improves the world for a rare species gets punished. The ESA has fallen into the black-or-white trap set by traditional conservation: Every place is either a wilderness reserve or a parking lot.

Maybe these facts would have upset me once. Today, they no longer look like a disaster, but like an opportunity tailor-made for reconciliation ecology. Even without all the scientific evidence that forces one to campaign for reconciliation, Michael Bean and Robert Bonnie understood perfectly. You can't win the war if you won't fight on the battlefield.

Whoops, sorry. War is the wrong image. We're trying to replace it with reconciliation. Perhaps I should have said, "You can't save woodpeckers in a cornfield." If private land is important, then we need to come up with a way to get private land involved.

Under the leadership of Bean, Bonnie, and their colleague David Wilcove, the Environmental Defense Fund designed a system to reverse all the pernicious ESA side effects it could. The USFWS adopted it in April, 1995 and called it "Safe Harbor."

Safe Harbor rewards responsible stewardship by private landowners. Suppose they go beyond the minimum requirements of ESA and improve the habitat they own on behalf of one or more T&E species. Then their property rights acquire complete protection from ESA restrictions. Of course, should they undo the good work they have done—their right under the law—they lose that protection.

Safe Harbor caught fire in the sand hills of the Carolinas, where landowners have enrolled over 20,000 acres. These properties range from horse farms to small forests, from residences to resorts. There are even some golf courses in the program. It is another thrust on behalf of longleaf pine and red-cockaded woodpeckers. One can easily imagine that many of the same methods developed at Eglin Air Force Base will be mustered into use on a much smaller scale. But Safe Harbor has even more significance than that.

Safe Harbor promises a new and enlightened future for the management of private lands. It encourages reconciling the use of land by humans with good stewardship on behalf of wild species—especially those species in trouble. People who do the best they can to heal the land and protect its species will, when they succeed, not be deprived of their right to use it for pleasure or profit.

The premise is simple. Whoever has improved the world for a rare species does not need to be convinced—let alone forced by some strict law—to do it. They are doing it already. Sounds stupid when I put it that way, doesn't it? But before Safe Harbor, whoever improved the world for a rare species got punished, hit with a quiverful of restraints.

Safe Harbor also promotes reconciliation in a social sense by luring landowners and private conservation organizations and government agencies into cooperating with each other. It does so by creating a game in which all stand to win by cooperating and agreeing. Play nice and you do better. Landowners are not stupid or evil people. They too would rather live in a rich world than a sterile one. So, they play nice.

Then hail to the Environmental Defense Fund, and to the clear thinkers at USFWS. The latter, especially, could easily have ignored Safe Harbor, but they did not. They adopted it, named it, and are making it their own. The civil servants of the USFWS must care about their mission. But I am not surprised. People who have made conservation biology their career are a passionate and dedicated lot.

CHAPTER 4

Making Money

> Even the law of gravitation
> would be brought into dispute
> were there a pecuniary interest involved.
>
> Thomas Babington Macaulay[1]

Will reconciliation cost money or make money? If I can show that it would make money, no doubt I would need to say little else to make it a wild success. Yet I have to begin the money chapter by insisting that this question must never be allowed to belittle people and their dreams. Yes, the money question is important. But I am certain that we worry about diversity for the best reasons: Aesthetics and Ethics. We think Nature is truly, stunningly beautiful; and we think it is our duty to care for her creatures. Disinterested yet passionate stewardship of the environment and its creatures may not be a religion. Nevertheless it invests human life with a sense of holiness. It declares with the unanswerable logic of beauty, that man counts in the universe.[2]

Some of us forget that, and peddle diversity as a cure for cancer. The idea is supposed to be that somewhere in the almost infinite soup of biochemicals that life contains, we will find magic bullets to conquer our diseases. But then what? Suppose, one day, we do identify all the arcane biochemicals that it takes to cure cancer (if, in fact, that is what it does take). What then? Should we throw away the useless species? Or maybe instead we should begin research on how grinding up the bones of some poor tropical bird will lead to better house paints?

Our forebears loved diversity long before they knew that diseases could be cured by chemicals. When first they fell in love with the sweep of a grassy plain or the rustle of a forest in autumn, they were taking all their ailments to witch doctors. When Adam and Eve gave each other flowers or marveled at a butterfly, they were trusting their aches to magic spells. Only their superstitions kept them from falling ill. But nothing kept them from thrilling at the sight of gazelles virtually flying across African savannahs. Our love for natural diversity has deep roots.

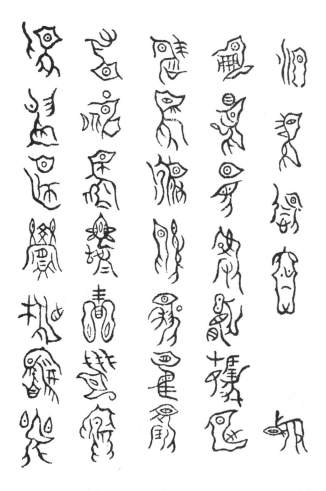

Oldest of all Chinese inscriptions—about 4,000 years old. It is a species list, enumerating game species. It is easy for me to imagine that the animal depicted by the lower right character is the brush-tailed Père David's deer, now extinct in the wild.

Many of my colleagues[3] are hard at work trying to find evidence that diversity delivers the goods. They suspect that ecosystems depend on natural diversity to purify our air and water. They hope that high natural diversities produce the greatest amount of food for us to eat and timber for us to use. They even believe it is possible that high natural diversities help protect us from massive outbreaks of disease. My colleagues have a long way to go. I wish them well.

But suppose they are correct. Suppose today's wetlands do purify our water supply and, in particular, do it better because they have a lot of species. Does that mean we should abandon wetlands as soon as some creative engineering firm discovers an artificial way to do the wetland job even better or cheaper? Does that mean we should then sell our wetlands for housing developments?

And what if my colleagues are wrong? What if diversity itself adds nothing measurable to ecosystem services?

Sure, we do continue to find new uses for old species. But God did not tell Noah to do that. God did not say, "Noah, old man, take every species with you on the boat. The day will come when you'll need one to clean up oil spills." God did not say that. Here's what God did say. God said "𝔒𝔣 𝔞𝔩𝔩 𝔱𝔥𝔞𝔱 𝔩𝔦𝔟𝔢𝔰, 𝔬𝔣 𝔞𝔩𝔩 𝔣𝔩𝔢𝔰𝔥, 𝔱𝔞𝔨𝔢 𝔱𝔴𝔬 𝔬𝔣 𝔢𝔞𝔠𝔥 𝔱𝔬 𝔨𝔢𝔢𝔭 𝔞𝔩𝔦𝔟𝔢 𝔴𝔦𝔱𝔥 𝔭𝔬𝔲."[4]

"*To keep alive with you.*" This is the commandment of reconciliation. "*To keep alive with you.*" Not in a zoo. Not even in those living museums, the preserved ancient ecosystem remnants that we cherish. No, God said, "𝔴𝔦𝔱𝔥 𝔭𝔬𝔲." Not only that, but "keep alive" is a perfect translation of the original Hebrew. The Hebrew uses a Semitic construction called "causative." Hence, "keep alive" means that we are required to actively cause all those species to stay alive.

And God charged Noah further, saying, "𝔅𝔯𝔦𝔫𝔤 𝔬𝔲𝔱 𝔴𝔦𝔱𝔥 𝔭𝔬𝔲 𝔢𝔳𝔢𝔯𝔶 𝔩𝔦𝔟𝔦𝔫𝔤 𝔱𝔥𝔦𝔫𝔤 𝔬𝔣 𝔞𝔩𝔩 𝔣𝔩𝔢𝔰𝔥 𝔱𝔥𝔞𝔱 𝔦𝔰 𝔴𝔦𝔱𝔥 𝔭𝔬𝔲—𝔟𝔦𝔯𝔡𝔰, 𝔟𝔢𝔞𝔰𝔱𝔰, 𝔞𝔫𝔡 𝔢𝔳𝔢𝔯𝔶𝔱𝔥𝔦𝔫𝔤 𝔱𝔥𝔞𝔱 𝔠𝔯𝔢𝔢𝔭𝔰 𝔬𝔫 𝔱𝔥𝔢 𝔈𝔞𝔯𝔱𝔥. 𝔏𝔢𝔱 𝔱𝔥𝔢𝔪 𝔰𝔴𝔞𝔯𝔪 𝔬𝔫 𝔱𝔥𝔢 𝔈𝔞𝔯𝔱𝔥 𝔞𝔫𝔡 𝔟𝔢 𝔣𝔯𝔲𝔦𝔱𝔣𝔲𝔩 𝔞𝔫𝔡 𝔪𝔲𝔩𝔱𝔦𝔭𝔩𝔶."[5] I know that the part about our "having dominion" comes soon afterward. And I also know that some self-appointed Bible experts—not the least of which, George Perkins Marsh, that great, mid-nineteenth-century environmentalist who first championed the notion that people mold nature by their actions—encouraged mankind to fulfill its religious obligation by converting all nature to human uses.[6] Still others, undoubtedly giving Marsh far too much credit for influencing human behavior, have traced all our ecological woes to that "dominion" bit.[7]

Nonsense. "To have dominion" does not mean to ruin; it means to govern. And, like it or not, we do govern. We have subdued the Earth.

But the Bible does not ask us to destroy the Earth. On the contrary, the Bible commands us to let all Earth's creatures be "fruitful and multiply." In fact, even Marsh's own views matured and mellowed. In his monumental and stunningly influential treatise of 1864, Marsh[8] devoted an entire chapter to the "Transfer, Modification, and Extirpation of Vegetable and of Animal Species." In other words, Marsh began to worry about extinction and about the introduction of species to places where they do not naturally occur. Marsh began to worry about the threat humans pose to species diversity. Moreover, in his introduction to that volume, he warns us that "the earth was given to (man) for usufruct* alone, not for consumption, still less for profligate waste."

Those of us who are neither Jews nor Christians will tap their moral roots into a different inspiration. But which of those other teachings ignores nature? Which demands that we strip her of her wonders and rape her? Certainly not those of the Moslem or Hindu, whose traditions demand immense respect for Earth's creatures. Certainly not those of that veteran Algonquin people, the Abenaki, whose creation myth bubbles with love of natural beauty and empathy for our animal cousins. Our moral systems may not all be the same, but I am aware of none that advocates either destruction of nature or even indifference to her needs.

Never mind. This is meant to be a chapter about making money. A few of us seem wired up to love that more than anything else. Those are the very types who tend to control disproportionate amounts of the land's surface. If practicing reconciliation ecology on their property means they'll take a big financial hit, at least some of them will keep their ethics and aesthetics under very close control. Business, they will remind us, is business.

Exactly what is the question here? Is it, Can I make a good and fair profit while practicing reconciliation? Or, is it, Can I make the same or more profit while practicing reconciliation? I suppose that different folks won't all be asking the same one of those questions. And all those who ask the first certainly won't define good and fair profit the same way. So let's cut to the chase. Let's ask the toughest question. Are there examples of reconciliation causing increased profits?

Usufruct: a legal term meaning, according to the New Century Dictionary, "the right of enjoying all the advantages derivable from the use of something which belongs to another so far as is compatible with the substance of the thing not being destroyed or injured." Marsh was a lawyer.

From the Abenaki creation myth

In the beginning there was nothing, only the Creator, and He was everything. Because He was everything, all around Him was nothing. If He was still, nothing moved. If He was silent, there was nothing in the universe to hear.

The Creator sensed the emptiness all about, and at the core of His being He became lonely. And so it was that He decided to make creatures of every kind.

Grand and beautiful were the creatures that He made. Each one, he made their bodies from clay, the flesh of our mother the Earth. Then with his own breath, the Creator filled them with life.

When the sun rose on that first morning of creation time, the men and animals opened their eyes on a world so full of beauty that there were no shadows in any place, but everywhere the world was a rainbow of colors to be seen. So beautiful was our mother the Earth when she was young that we would be amazed.

The men and animals walked softly upon the Earth. They spoke reverently to one another saying: "Oh, look at the great beauty of this place and everything that is in it!"

When we come at last to see the endless beauty of this world, then we come to know the beauty that lies within human life and the endlessness of the human spirit indeed.

Seven Eyes Seven Legs
Supernatural Stories of the Abenaki
(pp. 45, 49, 50)
Gerard Rancourt Tsonakwa and Yolakia Wapitaska
(Kiva Publishing: Walnut, California, 2001)

Do you now expect a speech about ecotourism? Not from me. Ecotourism is a great idea, granted. It can make more money than destruction, especially over the long haul. But it is not reconciliation. Ecotourism is non-exploitative use of an archaic ecosystem. Its optimum venue is a pristine landscape. Ecotourism takes nothing but pictures.

Moreover, ecotourism is a solution that does not multiply. It can never take place on more than a minor fraction of the Earth's wildlands. There just aren't that many rich western ecotourists to milk. In chapter 8, you will learn that this proportion is crucial to diversity, so ecotourism will not be much help.

If not ecotourism, then what? We need clear cases of cash benefits to satisfy the most skeptical. They exist. I will describe three. A ranch in Utah, the high Andean puna of Peru, and an Israeli resort and hotel construction zone a couple of miles from the Red Sea Star Restaurant that I described in chapter 1. They are all doing well by doing good.

Deseret Land & Livestock Corporation[9]

Many religions own much more than their schools and places of worship. The Mormon church is no exception. Among its enterprises, it counts a 200,000-acre working cattle ranch near Ogden, Utah. The Deseret Land & Livestock Corporation (DLLC) raises about 7,000 head of cattle and 2,000 sheep on this ranch. The job of the DLLC is to make money by producing food. It is very good at its job.

The DLLC sits on some pretty varied real estate. Elevations run from about 6,000 to 8,500 feet. Vegetation includes forests, sagebrush, and semi-arid grassland. Not all of this suits cattle.

A substantial number of wild game species also live on the ranch, taking full advantage of the range of habitats it encompasses. Moose, elk, mule deer, and pronghorn antelope. Ducks, too.

In 1976, somebody had a brainstorm. They noticed that about 3,000 hunters per year invaded the ranch for free, pursuing a trophy or some venison. Why not manage the game species, too? Monitor their habitats and keep them healthy. Let them flourish. And, naturally, sell the right to hunt them, thus expanding the profitable operations of the ranch.

This has turned out to be a fine idea. It certainly benefitted the game species. DLLC cut way back on the number of hunters allowed on the ranch, and some game populations responded sharply. A recent census counts perhaps 100 moose, 5,000 deer, 2,000 elk, and 500 pronghorn. Elk, for instance, numbered only 350 before management.

The idea to manage the game has also helped many nongame species. These other species share the ranch's habitats with the cattle, the game animals, and of course, the people themselves. Mink, jackrabbit, and squirrel. Beaver, bear, bobcat, and puma. Sage grouse and other birds. Improving game habitat also improved theirs as a by-product.

The DLLC set up its own Department of Wildlife Management to care for the game habitats, keep track of game populations, market the hunting privileges, and harvest the cash. Every year, DLLC makes 25 to 50 percent of its profits from its Department of Wildlife Management. In 1994 alone, hunters forked over $565,300 in hunting fees. In addition, the conservation uses of the ranch create jobs. DLLC pays wildlife biologists and graduate students, cooks and guides and trail-packers.

The Vicuña of the Puna

A vicuña coat. It was a princely bribe, worthy of the highest placed, most powerful confidant of the president of the United States. A coat made of the wool of the wild, graceful, humpless camel called the *vicuña* that roams the *puna* (the cold grasslands of the high Andes). Such a coat helped to bring down Yankee blue blood, Sherman Adams, member of the Henry Adams family of Quincy, Massachusetts, a family as close to nobility as we Americans have ever known. The Adams family has produced two presidents, scholars, statesmen, and timeless authors. Until the disgrace of his vicuña coat, Sherman Adams himself marched solidly in their tradition. Sherman Adams served as governor of New Hampshire from 1949 to 1953. Then he got the choicest appointment one can imagine. Adams, respected adviser to Dwight David Eisenhower, became, in 1953, the first chief of staff of any U.S. president.

But Adams's career disintegrated in 1958. He resigned for accepting a vicuña coat from some businessmen interested in illegitimate leverage at the White House. Sherman Adams had crossed the line that Americans have guarded for over two centuries. He had behaved like a prince instead of a respected servant of democracy.

A vicuña coat. The history books will tell you that there was also an oriental carpet involved, but I remember this scandal. No one talked about carpets, just coats. Everybody had carpets. But nobody had ever before heard of a vicuña coat.

A pair of vicuña in Peru's Pampa Galeras National Reserve. © Tui De Roy, Roving Tortoise Worldwide Nature Photography.

Vicuña fleece is softer, lighter, and warmer than the finest cashmere. And vicuña fleece is rare—and very expensive! In 1998 dollars, raw vicuña fleece brought about $500 per kilo. A vicuña scarf sold for about $400, a sweater for $3,000, and a coat for $20,000. Vicuña fleece is the world's most luxurious fiber.

In pre-Columbian Peru, no one but Inca nobles was allowed to wear vicuña fleece—others caught doing so were executed.[10] To obtain the fleece, the Incas participated in *chakus:* They organized themselves into large teams that herded the vicuñas into special pens where shearing took place. Of course, after shearing, the vicuñas were set free to grow a renewed crop of wool. The vicuñas were precious and carefully safeguarded.

But the Spanish colonists did not bother with shearing—at least not until they had first shot the vicuña. After all, vicuñas swarmed over the puna. There were about 2,000,000 in 1492. Who would miss the ones that got shot?

Nevertheless, by 1820, when independent Peru was born, the vicuña was already in trouble. Peru prohibited their slaughter, but the law was ignored. The Peruvian vicuña population continued to decline. About the time Mr. Adams accepted his coat, it had dropped to about 10,000. By the late 1960s, Peru may have had as few as 5,000 vicuñas, and only about 2,000 survived in all the rest of the species' range (Argentina, Chile, and Bolivia).[11]

In 1975, an international treaty banned the sale of vicuña fleece all over the world. Without a doubt that treaty rescued the species from extinction. In Chile, where good annual censuses began immediately, vicuña populations grew from 2,000 in 1975 to a rather steady 20,000 by

1992. Ecologists estimate that 20,000 or perhaps 25,000 is about all that the Chilean range can support.[12] Because of the recovery, vicuña trade was re-established in 1987, but it was limited to the healthier populations and to fleece taken from live animals.

Finally, in 1991, Peru transferred the custody of the vicuñas to some 600 mountain villages scattered about the puna where the vicuña live. The old Incan techniques of corralling and shearing were reintroduced.

The vicuña are now thriving and so is trade in their wool. The growing population of vicuña now exceeds 120,000 in Peru plus 60,000 in Argentina, Bolivia, and Chile combined. Their fleece already brings in over $600,000 per year to the Peruvian villagers who run the chakus. That's not the same as finding an oil field on your land, but $600,000 per year is profit, and to these mountain villages, it represents a unique opportunity to improve their communities' infrastructure.

In my home, I keep a relic from my childhood—*The Story Book of Wool*. It's a small book, part of a set by Maud and Miska Petersham that I have kept—decorated with my superfluous crayon marks—for my grandchildren. Toward the end, it actually depicts some vicuñas in their high Andean habitat. It accurately reveals the high cost of their wool, the rules enforced by the Inca, and the conservation efforts of Peru. But it also states, "It is necessary to kill the animal in order to obtain its fleece."[13]

Remarkable. How easy it is for us to glide, unaware, from what is commonly done to what must be done. In 1939, when the Petersham book was first copyrighted, it had been centuries since anyone had harvested vicuña wool without slaughter. The memory that it once had been taken only from living animals was almost lost. May the vicuñas teach us that what *is* rarely scratches the surface of what *could be*. As we look around the habitats of our world today, let us appreciate more and more that we've hardly seen anything yet.

Eilat's Hotel District and Its Salt Marsh

Until you've had a look at the rest of the region, you'll never understand why a multitude of birds should care about the salt marsh of Eilat, Israel's southernmost city. But once you have toured the surrounding desert, it is impossible not to. If you are a bird, Eilat's salt marsh is the only place to forage for a hundred miles around.

Eilat rests in the great Syrian-African Rift Valley along one of the most heavily used bird-migration routes in the world. A finger of the Red Sea—jabbing northeastward through bleak deserts in Egypt, Saudi Arabia, and Jordan—pokes at Eilat's waterfront. To the north, the floor of the rift valley climbs quickly out of the city into more bleak deserts along the Israel-Jordan border. From the valley floor, the walls of the rift rise steeply and bleakly toward rocky, arid highlands that are bleaker still.

Bleak. Bleak. Bleak. Nowhere does the average annual rainfall exceed one inch. What's a poor bird to eat? All that traveling, stomach on empty, and there's nothing to restore the soul except the little 12-square-kilometer salt marsh of Eilat. Without that salt marsh and its buzzing breakfasts, a crucial link in the rift-valley migration route would be broken.[14]

And the route is very heavily traveled. A substantial portion—as much as a third—of the birds of Europe and Western Asia use it to get back and forth from their African wintering grounds. Up to 98 percent of the individuals of some European species depend on it as well. Approximately one billion birds pass through every year. Observers have recorded 257 species so far. No one knows how many of them would vanish were Eilat's salt marsh to disappear.

But Eilat has another face, a human face. Eilat is a great resort city. In wintertime, multitudes of pale Europeans stumble out of Airbusses into Eilat's warm sun and head for its coral-fringed beaches. No doubt those with a nose for such things can actually smell the money these tourists bring along.

Eilat has become a boomtown. Massive, luxurious hotels. Nightlife. A boardwalk. Fancy and not so fancy shops. Greasy and not so greasy restaurants. Build, build. It's birds versus builders. But you know who wins. Indeed, the salt marsh is nearly gone. Nonetheless, this story has a happy ending. Reconciliation ecology averted the evil decree.

The hero is Dr. Reuven Yosef, director of the International Birdwatching Center of Eilat. Yosef knew about a tract of wasteland—a garbage dump, really—that absolutely nobody wanted for a hotel. From Yosef's perspective, however, it was first rate. It was right next to the last remaining acres of salt marsh. Low lying and shrubby desert, yes. Full of refuse, yes. It was not a salt marsh and could not support the migration of so many songbirds, but to Yosef it looked like the promised land.

Yosef had a vision. Put a few hundred tons of landfill here and there, scrape it into dikes, cut a channel to allow some seawater in, plant some of the right species of shrubs and *voila!*, you would have an artificial salt marsh complete with open water areas. A rich oasis for songbirds, wading birds and floating birds, too. Unfortunately, all that costs money. Where would it come from?

Yosef noticed a peculiar thing. As the construction shovels excavated the hotel sites, they loaded the dirt into dump trucks. The dump trucks soon disappeared up the hill toward the Dead Sea. When they returned later, they were empty. Where was the dirt going? Could he get it for his dream?

Yosef found out that the dump trucks operated under a city ordinance. Whoever dug a big hole in Eilat had to transport its contents up the hill, far away into the desert where tourists wouldn't see it. That costs money, too. But it was the law.

Opportunity was knocking. Yosef's dream tract was just down the road from the builders' construction sites. Maybe the city would permit them to dump the dirt where it would do some good? Why not? The builders would save. The birds would be saved.

A deal was reached: Builders did not have to transport their dirt so far if they would help Yosef put together the artificial salt marsh. The builders increased their profit and Yosef's vision became a reality.

A view of Reuven Yosef's pseudo-salt marsh, Eilat, Israel. On the horizon is part of the swarm of tourist hotels. Just beyond them is the ocean, the Red Sea Star Restaurant and the coral reefs of the Red Sea. © 2002, Evolutionary Ecology Ltd.

Today's patch of salt marsh little resembles its predecessor or even any natural habitat. It is carefully built up, contoured and planted on a refuse dump. Many of its plants never grew in the Eilat area before. It is regularly irrigated with treated, nutrient-rich sewage water. But I have seen it do its job during spring migration. It produces food for wildlife at four times the rate of the natural salt marsh that preceded it. It works.

What Are the Odds?

How exceptional are the cases that I described? Will reconciliation ecology often lead to higher profits? The Eilat salt marsh is pretty special, after all. That is the very reason for replacing it. So are vicuñas. And if every ranch and orchard had exciting game to hunt, the profit to be made from hunters would quickly get spread too thin to pay for the costs of managing the game populations.

To firm up the case for the profitability of reconciliation, we need to look at the intrinsic money-making practices of the land we use. We must ask the farmer if the farm would make as much money from its crops if it were managed using reconciliation ecology. We must ask the timber company if the forest would make as much money from its wood if it were managed using reconciliation ecology. We must ask much the same question about the ranch, the pasture, the orchard. When the dust has settled, when the posturing and false examples have been exposed, when confusion has been defeated, those will be the questions that count.

Considerable hope comes from experiments being done on two grain-growing farms.[15] One of these farms sits on some of the richest farmland in Pennsylvania, the sloping soil of Berks County in the southeastern part of the state. The other spreads over a tract of east-central Nebraska with much deeper soil, but just barely enough rainfall to avoid needing to irrigate the crops. In each of these cases, the farms have experimented with various regimens of plowing and chemical use. Some fields were treated with the best intensive agricultural methods, including very high levels of chemical fertilizers and pesticides. Others received no chemical fertilizer, or no pesticides, or shallower plowing, or a combination of these.

Experiments began at the Rodale Research Center, Kutztown, Pennsylvania, in 1981. Fifteen acres were set aside and farmed without chemical fertilizer or pesticides. Instead of the usual mix of corn and soybeans, this acreage grew a mixture of grains—corn, soybeans, wheat and barley. Instead of chemical fertilizer, they plowed animal manure or legume manure into the land. Some of the land was plowed conventionally, some much less deeply.

The idea? A large portion of a farm's chemical fertilizer washes away and pollutes waters. Deeply tilled soil invites serious erosion. And no one doubts that the environment can be harmed by pesticides. If farmers could avoid such environmental loads, they would do themselves and the nation a big favor. But could they do it and still make money? Could they do it and still feed us?

Once the new methods got going (it took several years of transition), they yielded plenty of food—in every case, more than the chemical methods did. Even if we ignore the extra yield of wheat and barley, they actually produced as much or more corn and soybeans than the environmentally damaging chemical methods! And they were even more profitable. The environmentally friendly methods all brought in about $48 per acre per year (net); the chemical methods varied from less than $6 to $44 per acre per year. (Furthermore, I have not mentioned the hidden costs of soil loss. As we will see in the next chapter, these losses make the chemical methods seem virtually useless for food or profit except in the very short term.)

In Nebraska, the story turned out to be different. Here, on land of the University of Nebraska, a corn-and-beans farm outperformed all others regardless of whether it was treated organically or with herbicides and chemical fertilizer. Surprisingly, the addition of the herbicides did nothing to help profits, but the organic methods did depress them a very small amount. Corn and beans grown only with chemical fertilizer yielded $125 per acre per year (net), whereas corn and beans grown organically yielded $118. In addition, although yields of soybeans stayed the same, corn yields declined about 30 percent with organic methods. (Again, I warn you that these figures do not include the cost of soil loss.)

O.K. In Nebraska, the chemicals win. Organic methods reduce net profit 5.6 percent. Nevertheless, the organic farmer still makes a handsome profit. How important is it to squeeze that last penny from the Earth?

Extreme farm methods have plenty to do with species diversity. A landscape poisoned with herbicides and pesticides might as well be an asphalt parking lot for all the good it will do most native species. Land so abused has been wrenched away from diversity by a domineering tyrant. We need instead to become benevolent sharers of our world. If we can learn to do without biocides, we shall at least have a chance to develop farming methods that reconcile crop growing with diversity preservation.

As a matter of fact, reconciliation itself may help us do without the poisons. Many studies suggest that reconciliation has considerable potential for minimizing losses to agricultural pests. A classic case concerns the leafhopper pests of California vineyards.[16] The grape leafhopper chews up the leaves of grapevines very badly. Sometimes there are so many leafhoppers that the grapes are ruined and the profit turns to a loss. Often, pesticides either fail to help or actually make matters worse.

A certain species of native wasp thrives on grape leafhoppers. It lays its eggs inside the leafhoppers. The wasp larvae then destroy their prey from the inside out. Result: biological control. But there is a problem. The wasp fails in some vineyards. Why?

Doutt and Nataka solved the puzzle. Grapevines shed their leaves in autumn. So, in the winter, the grape leafhoppers have nothing to eat and become inactive. The wasps get left out in the cold, and die.

But the wasps do not become extinct. They just disappear from the vineyards. Next year, they must establish themselves anew. That gives the leafhoppers a head start and dooms the effort at control. So, how could the wasps ever succeed?

The wasps succeed wherever the vineyards grow near patches of native blackberry bushes. Native blackberries are not deciduous. They harbor a different leafhopper and it continues to feed throughout the winter. The wasps survive by living on blackberry leafhoppers.

Grape growers took action. They planted patches of the native blackberries in shady spots near their vineyards. Wasps live there and quickly reinvade the vineyards in the spring, thus keeping the pest populations down.

The growers are practicing reconciliation ecology on a small scale, expressly in order to enhance their profits. They studied the biology of a native wasp species and designed a mosaic of habitats to encourage its population. The mosaic includes patches of native blackberry and harbors all the native species associated with it. Granted, their goal was not directed toward preserving biodiversity, but it might as well have been. It achieves precisely that.

Much of the land that people use is devoted to agriculture, so we ought to be quite cheered by the numerous examples that show how agriculture can help protect diversity and, simultaneously, make a good profit from

crops. Sometimes these compatible agricultural uses occur quite accidentally, but in other cases they are deliberate. They exist in pasture-
lands, croplands, plantations, and timberlands. They come from rich and
poor countries and are sponsored by private by governmental agencies.
Cardamom-growing is one nice case because it comes from a traditional
crop grown in plantations but without the financial resources of a wealthy
Western agribusiness.

Cardamom, the dried berry of an herb (*Elettaria cardamomum*), is a
valuable spice native to the evergreen tropical rainforests of southern
India. Growers cultivate it there in large plantations covering some 810
square kilometers. Cardamom grows only in the shade of taller plants
and also depends on cross-pollination by insects, principally honeybees.
Growers maintain many tree species in their plantations. They do so
both to provide shade for cardamom and to maintain a steady supply of
nectar for cardamom's pollinators. Bees visit 37 tree species growing in
the plantations. Ten of the thirty-seven supply both nectar and pollen,
three supply nectar only and the rest supply pollen only. However, from
May to September, flowers are not very abundant, and pollination rates
suffer. Biologists, working entirely from the perspective of the growers,
want to increase plant diversity by adding species that do flower during
these months. Such plants would bridge the temporal gap and provide a
steadier nectar supply for the bees. Hence, such plants would increase
both cardamom production and growers' profits.[17]

Quite a few investigators see promise in traditional methods of agriculture. For instance, the mixed gardens of Java raise 607 species of plants,
yet about as many insect species live in them as we would expect to see
in a deciduous subtropical forest.[18]

But do the native species actually maintain self-supporting populations in the Javanese gardens? Maybe the diversity just leaks in from
neighboring natural systems? We do not know. And we also do not know
whether an enterprising, swashbuckling agribusiness corporation could
figure out how to make even more money from the land. (We do know,
however, that where agribusinesses have done such things, their calculations often do not hold up for more than a short time. On the other
hand, the traditional systems are sustainable. See the example of the new
coffee plantations in the next chapter.)

So, reconciliation ecology doesn't have all the answers yet. Perhaps
when we get them, we will learn that saving the species of our world

won't be entirely free. But isn't beauty worth *something*? Jane McAlister Pope does not doubt it:

> True beauty adds a quality to our lives that goes beyond the aesthetic. It's about joy, not mere prettiness. Beauty cuts a direct path to the soul. No human artifice can compare to the simple wonder of creation.[19]

And wouldn't you be willing to pay a modest price to keep your moral precepts? Good, you'll get a two-for-one special. Aesthetics and ethics. What a deal!

Beware, though. Our answers will depend on the system we use to calculate the profits. While we await the numbers to analyze, let's encourage our children—for they will be the ones to compute the net costs—to keep their wits about them. Let's hope they stay clear of con games. Let's hope they are not fooled into paying part of the costs themselves, or worse, foisting them off onto their defenseless, unborn children. They need to make sure the whole business operates according to a rational system of rewards. They ought to be fair, to make sure the costs are borne by society as a whole, and not the individual landowner. And they must not neglect the social costs of doing business one way rather than another. In the next chapter, I will illustrate how important all these points are.

CHAPTER 5

Hidden Costs

You aren't too good with the truth, either,
your species. . . . I can see there might be a
positive side to this wilful averting of the eye:
ignoring the bad things makes it easier for you
to carry on. But ignoring the bad things makes
you end up believing that bad things never happen. You are always surprised by them. It surprises you that guns kill, that money corrupts,
that snow falls in winter. Such naïvety can be
charming; alas, it can also be perilous.

Julian Barnes[1]

When a skillful graduate of the Wharton School tackles the job of projecting profits for her agribusiness, she ignores lots of things. No, she is not trying to dodge taxes. She is simply doing her job well. Her company's profits depend entirely on the money actually paid out or received. In contrast, those profits have nothing to do with various hidden environmental subsidies. She may not even be aware of the subsidies, and her company may not be either. She does not worry about these subsidies only because her outfit never gets billed for them. Nevertheless, they are as real as depreciation.

Let's look at just one of these environmental subsidies: the soil gift.

Agriculture thrives on the rich, deep topsoils of mid-North America. Those soils grow grains aplenty and feed us more cheaply than any other

WIN-WIN ECOLOGY

society in the history of our species. We revel in their quality. We farm them as intensely as we know how, always with a relentless drive to improve their yield and our efficiency. We export their superabundance to peoples rooted to less productive soils. How fortunate we are.

But our good fortune is a space *and time* phenomenon. Two million years ago, these soils were significantly shallower and poorer. The soil built up as a result of the glacial activity of the last ice age. Cooler temperatures slowed the rate at which nutrients decay. Grinding ice sheets pulverized the bedrock and speeded its conversion to earth. We are not just lucky to live in North America. We are lucky to live in North America during the twenty-first century.

Wes Jackson, having reviewed the scientific literature, reports that it takes nature about 3 to 10 centuries to build up a single inch of top soil.[2] But our agricultural practices wash it away at the net rate of six inches per century. Even with the help of added nutrients and the most sophisticated soil science, we cannot force the topsoil to build up at more than half the rate we wash it away.

So far, massive amounts of topsoil have washed into our waters. To fight those losses, the people of the United States have spent tens of billions on the U.S. Soil Conservation Service (SCS) since its founding in 1935. Yet despite the best efforts of the SCS, our topsoils still get thinner every year. (I mean no criticism. Heaven knows how much worse it would be without their efforts!)

The U.S. government estimates[2] that by 1979, we had virtually destroyed about 100 million acres of cropland. That is one out of every six cropland acres in the area of the continental United States.

It does not take an agricultural economist to realize that such destruction cannot be cost free. Yet we make no charge for it, so it does not appear in agribusiness balance sheets. Farmers pay good cash for farmland, and they pay in proportion to the depth and quality of the soil. As they depreciate that soil, should they not have to account for it? Is accounting theory so unimaginative that no one can figure out a way to keep meaningful records?

Remember those farming experiments in Pennsylvania and Nebraska? Recall that, in Pennsylvania, farming without chemical additives brought in the most food and profit, while in Nebraska, it brought 5.6 percent less profit. But, as I warned you, we did not include soil losses in our reckoning. Now we will take them into account.[3]

56

The picture in Nebraska changes only a little when soil loss is included. Economists figure the Nebraska soil loss from chemical farming methods is worth about $5 per acre per year. However, farming organically costs no soil at all. So farming organically does not reduce profit 5.6 percent, but only 1.9 percent. The chemical way still wins—but only by a nose. More important, think about the long-term difference. Soil loss repeated over enough generations ruins all hope of profit.

In Pennsylvania, meanwhile, consideration of soil losses turns Nebraska's leak into a tragic torrent. In Pennsylvania, the maximum net profit with chemical additives goes from plus $44 per acre per year to *minus* $22. In Pennsylvania, a conventionally tilled, chemically treated cornfield *loses* $87 per acre per year. In Pennsylvania, chemical farming may not have seemed as profitable as nonchemical methods, but at least it seemed to return some black ink. Now, when we pay attention to the soil losses, we discover that the black ink was actually red.

Even with nonchemical farming, Pennsylvania farmers still lose some soil. But not much. Careful shallow tilling held the soil loss to less than $10 per acre per year. Every one of the nonchemical methods turned a real profit. The best one made $38 per acre per year, not much less than before we subtracted the soil losses.

This is not a book on environmental economics, so I won't review the cost of cleaning up the water that carries away our soil. And I won't detail the environmental costs of the air pollution created by burning coal and oil to maintain the land in production. But you had better believe we taxpayers do pay for cleaning up that water and that air, too.

What is my point? Not the details. Norman Myers has already told us the details of hidden subsidies,[4] and we ought to listen. My point is that hiding the details of our subsidies makes comparisons irrelevant. We must have those details to compare the profits from sustainable, reconciled agriculture with those of destructive, modern agriculture. The profit and yield calculations of extractive businesses must change to reveal these hidden subsidies. Whether we foist off our costs on our children's children or pay as we go, someone eventually will have to pay.

Society has no long-term interest in giving away the food factory. In particular, it has even less interest in giving it away to those whom we force—with our negligent accounting methods—to ruin it.

Our forests tell much the same story. Consider the traditional way the U.S. Forest Service calculated its profit from selling timber on the 192 million acres it administers for the people of America. It ignored 99 percent of the costs of building the logging roads necessary to harvest the timber. Only a penny out of every dollar was reckoned in the balance sheet. Keep in mind that the Forest Service builds more miles of road every year than any other federal or state agency! Each square mile of timber-yielding forest in the Pacific Northwest requires five miles of logging road.[5] Who can be surprised that by virtually ignoring this major cost, the Forest Service always showed a net profit on its timber sales?

The balance sheet changes dramatically as soon as the road costs enter the calculation. In 1997, the first year the Forest Service included all their road building costs, it had to admit that it lost $88,600,000. Could it have done better with a reconciled scheme for managing our forests? We do not have such a scheme, so we do not yet know.

Isn't it time to learn how to do things better? After all, by law, the Forest Service must also manage our forests to protect biodiversity. If Eglin Air Force Base can sell its timber at a profit while protecting biodiversity, why not the U.S. Forest Service?

Irrational Systems of Rewards

In chapter 3, we saw the self-defeating way the Endangered Species Act once treated the good private landowner. Before the Safe Harbor program, the ESA punished those who successfully cared for their land. If a rare species rewarded their stewardship efforts by moving on to their property, the ESA punished the steward by imposing restrictions on what such responsible landowners could henceforth do with that property. Meanwhile, the ESA ignored the landowner who destroyed habitat—habitat that someday might well have become important to a rare species. Oppress your friends, reward your enemies. A pernicious, irrational system. What a pity!

Thankfully, Fish & Wildlife's Safe Harbor program has changed the system. Now the landowner has incentives to practice both ethics and aesthetics. Do the right thing, and your property rights get protected. Safe Harbor adds to the value of your land. A rational system makes all the difference.

For a second example of an irrational system, we turn to the coffee plantations of Latin America.[6] Brought to the Americas in 1726, coffee is the fruit of a native African shrub adapted to grow in the shade of tropical forests. In a traditional coffee plantation, the shade comes from various species of tropical trees growing rather densely. For instance, a traditional coffee plantation near Managua, Nicaragua, would have at least 25 species of trees. Often, the trees generate something of value in addition to shade: bananas, citrus, guava, firewood, or timber. Some censuses have counted anywhere from 120 to 561 trees per hectare of coffee plantation.

The grower tends each coffee plant carefully. He prunes it with skill, reducing the shade it casts and allowing him to plant small amounts of light-loving annual crops such as yams. When the coffee plants have grown old, the grower replaces them with smaller, younger individuals. This creates an even larger light gap, in which the grower often plants corn or beans.

The traditional coffee plantation harbors a lot of diversity. Not all the species of the pristine forest can live in it, but most can. The managed-for-coffee forests of the Maya in Tamaulipas have more than 300 species of plants. Shade-coffee plantations are rich in vertebrates including birds, snakes, bats, and other mammals. In addition, diversities of beetles, spiders, ants, and wasps compare well with those of some of the region's natural habitats. Shade-coffee plantations even produce loads of leaf litter, giving life to the forest floor and so supporting a wide variety of soil insects and other species. Vandermeer and Perfecto estimate that the traditional coffee plantation preserves 60 percent to 70 percent of the biodiversity of the wild tropical forest that it replaced.[7] No one has yet done the work to find out what fraction of these species have sustainable populations in the coffee "forests." But evidence already suggests that, when natural forests were at a low ebb in Puerto Rico, traditional coffee plantations there saved many rare orchid species from extinction and minimized the loss of birds.

Traditional coffee plantations are reconciled habitats. They combine food-growing and profit with habitat for wild things. Traditional coffee plantations even pass the test of being "new." We invented them, albeit centuries ago.

In 1970, spores of an ominous fungal invader wafted onto the shores of eastern Brazil. Called *Hemileia vastatrix*, the coffee leaf rust, it is a virulent destroyer of coffee plantations. During the last century, its depredations stopped coffee growing entirely in India and Sri Lanka. For tropical American coffee growers, the sky seemed ready to fall.

The U.S. Agency for International Development (USAID) suggested a technological war on coffee rust. Find varieties of coffee that grow well in full sun, because the sun's heat would deter the fungus. Pump in fertilizer to raise productivity. Pesticides would also be good. Cut down the trees. Rip out the other plants. Plant the coffee bushes in dense, neat rows so you can use efficient, mechanized methods to care for your coffee plants, techno-coffee plants.

I must share with you my skepticism at such hype. I have seen it before. It usually involves the combination of

- A new product or technology

- A sophisticated sales campaign

- Sleight-of-hand calculations that generally assume the new technology, when adopted widely, will not change the environment

- Manipulating the human herd instinct by suggesting that the new must be good because others are using it

- Scaring the dickens out of us

Sales campaigns to sell new technology drove streetcars and interurban rail systems out of existence. Now we wish we had not bought quite so many cars and busses, and we are paying dearly to reacquire what we used to have. Sales campaigns ruined our railroads. They convinced us to buy tasteless tomatoes and insipid beer. They even got us to wear polyester suits. Sophisticated sales campaigns do not always peddle crap, but whatever they peddle, we buy, at least for a time.

Many signs suggest that the new, techno-coffee plantation is a creation of an industrial ad campaign. Growers must buy new machinery to grow it. Techno-coffee growers need a lot of fertilizer and pesticides that they did not need with the old coffee. It costs about 6.5 times as much to grow techno-coffee as traditional coffee. Finally, coffee leaf rust turned out to be not so bad a problem in this hemisphere. Ironically, one of the few varieties of coffee that does have trouble with the rust is *caturra* coffee, a techno-coffee strain that was touted to be resistant. You see why I'm suspicious?

The campaign to promote techno-coffee worked. Today, from Mexico to northern South America, more than 41 percent of the 6.9 million coffee-growing acres use the techno-coffee system.

Whether or not techno-coffee is a huge economic and gastronomic success, its farming methods really knock the hell out of diversity. It begins by cutting down all the forest trees. Whoops, there go all the species associated with them. In addition, many species that live among traditional coffee plants, cannot live with techno-coffee. The techno-coffee plantation supports only about 5 percent of the species that would thrive in a wild tropical forest.

Here is the other side of the story: "The yields per acre go up. Acreage on techno-coffee plantations yields almost five times as much coffee as traditional methods. Isn't that better for the environment?" we are lectured. "We can grow more food on less land. There will be more for reserves."

Sure, and the check is in the mail. Once the sales campaign takes hold, the real problem becomes the reward system. Here's what happened as a result of techno-coffee's higher productivity: Land was not taken out of production to compensate for the productivity increase. Coffee production spiraled upward. In 1989, the International Coffee Agreement—designed to control coffee prices—collapsed. Without it, coffee glutted the markets and prices plummeted.

Twenty-eight coffee-growing nations responded by forming a new cartel in 1993. They signed on to the Coffee Retention Plan, in which members withheld 20 percent of their coffee from the market. It worked. Supply dwindled and coffee prices began rising. Of course, there was also a terrific cold snap in Brazil that knocked the 1994 coffee crop way down. But let's give the cartel due credit too for its economic bravery.

What is the status of Latin American coffee today? Worse than before. The most telling figure is the proportional contribution of coffee to exports. This value is declining. In Central America it dropped to 18 percent in 1993—from 22.2 to 41.2 percent in the 25 years from 1965 to 1990. In Mexico, it fell below 1 percent, when it had been 5.9 and 6.8 percent in 1965 and 1975, respectively.

Of course, many other economic changes have taken place in Latin America, especially in Mexico. The value of her coffee has been partly overshadowed by that of her manufactured products. That is believed to account for much of coffee's decline in importance to Mexico, but it does not explain the declines in other Central American nations, where coffee has experienced a substantial loss of export share in every case.

Even in the face of the astonishing new popularity of coffee houses in the United States, techno-coffee is not a clear success. The plantations produce more coffee, but earn less profit. I wonder if some of that profit could be leaking out in payments to North American chemical companies and machinery manufacturers? Maybe the Europeans are getting a share, too. If they are, they ought to give a commission to the well-intentioned guys at USAID. They started it all. Who says foreign aid is a dead loss?

Let's summarize. A benign method—maybe even a beneficial method—of profitable coffee-growing gets replaced by techno-coffee farming, which can grow about five times as much coffee per acre. But techno-coffee farming destroys most of the diversity, produces too much coffee, erases the profits, and causes the creation of an international cartel to keep the surplus coffee off the market. Insane, isn't it? We were surely better off before.

Why did this happen? Basically, the system determined the outcome.

Imagine that you are a coffee grower. You have your acreage to work and somebody shows you how to grow five times as much on it. You adopt the new method, calculating just how much more you will earn each year. You do that using the coffee prices you have been getting in the recent past. That is called the fallacy of extrapolation: You fail to ask yourself what will happen to those prices once all your competitors adopt techno-coffee and greatly increase the total amount of coffee on the world market.

Soon you are growing more, but getting less per pound. But you cannot stop. If you went back to the old way, your coffee production would decline. You would have to sell it at the same deflated market price and your family would starve. You are hooked.

Now you look across the road at your neighbor. He did not switch to techno-coffee plants. His family was happy with life as it was. Now they are destitute and he is desperate. His crops are bringing only one half (or maybe one third) what they used to. If it's not too late—if he has any credit left at all—he has to change. Maybe his friends can help him buy in. He wants to buy in as soon as he can.

Result: coffee production mushrooms. Prices decay. Diversity does, too.

Into the breach steps the cartel. They keep 20 percent of the coffee off the market and prices rise. But they cannot keep growers from producing it. The more each grower produces, the more money he makes because the cartel is required to buy the surplus.

Is there no way out?

An authoritarian cartel could, theoretically, do the trick by outlawing techno-coffee. But everyone knows that would be impossible in practice. All the governments in the cartel would have to keep their agreements to each other in perpetuity and in perfect faith. I don't think it would last five years. Since the 1960s, when coffee-growing nations first got together to manipulate coffee prices, they have failed to achieve anything very useful. Remember, it was the disastrous 1989 breakdown of their agreement that triggered the 1990–1993 crisis in coffee prices.

The answer will lie in changing the system. Governments have already intervened. Why not use that intervention to support traditional growers? They are not contributing to the problem. Why not make sure that traditional growers get the profits from traditional coffee that they would have gotten from planting techno-coffee? And why not remove support from techno-coffee plantations?

We must be skeptical when we calculate the profits of an unreconciled agricultural system. We have to make sure it really delivers the benefits it boasts of. If not, we have to change the system of accounting to give credit where it belongs and charge expenses to the true costs.

Social Costs

The family farm. One of America's great icons. We are often told that our American character sprang up from the family farm. Maybe so. But, without a doubt, the institution of the family farm suffered terribly in the twentieth century.

What is the family farm worth? What if reconciliation turns out to need family farms, but they turn out not to be quite so profitable as Behemoth Grain Incorporated? Should we write down zero where the balance sheet asks us to fill in the extra value of the family farm?

I do not have the training to address this question and will leave it to those who do. But I will find it virtually impossible to believe any economist who claims that family farms are not worth much. You see, I do have an inkling of what such farms mean because of my own experiences as a child in eastern Pennsylvania. Family farms offer a profoundly different world to those lucky enough to grow up among them. Economic balance sheets may resist quantifying their cultural wealth and values, but that tells you more about economic balance sheets than it does about the rich milieu of the family farm. So, in lieu of economics, I offer you a few of my memories following this chapter. That won't substitute for a scientific approach, but it will give you a notion of just how important and life-shaping and cherished such experiences can be.

So don't be afraid to disagree with cold economic calculations. True enough, aesthetics and ethics are for people with enough to eat, a place to sleep, and a covering for their bodies. If reconciliation ecology won't let you pay the bills, it won't do anybody much good. But don't get greedy. You don't have to be as rich as Rockefeller before you give yourself permission to love the flash of that bluebird zipping by. And you don't have to be Ted Turner before you are allowed to act morally toward God's creatures. To be truly happy, learn to value the intangible and the inestimable.

Interlude

A Personal Witness

I never lived on a family farm myself. The closest I've been is summers, growing a victory garden with my father between two family farms in eastern Pennsylvania—Harleysville. Pennsylvania Dutch folks owned and worked those places. That sonufagun Harry Truman was the president of the United States.

The neighbors' farms were mixed farms. Both grew corn. One also had pigs, the other, dairy cows. Didn't look very romantic to me. Those neighbors worked long and hard.

The pigs stank pretty bad, so we kids usually kept our distance. But we loved to sneak into Durstine's cow barn and dive into the haystack from the top of the rafters. Once, Mr. Durstine chased us away with a pitchfork and told us never do it again. But he didn't say anything about his insurance policy. I think he was just afraid we would get hurt.

Toward the end of August, when the ears got ripe, Mr. Durstine would let us have a corn boil. We'd set up a big fire at the edge of the corn field and hang a cauldron of water over it. After the water boiled, we'd go into the field and pick some corn. Then we'd run back to the pot, shuck the corn and dump it in.

I'm not kidding. We'd run as fast as we could. That pot was only 50 yards away, but every second counted. Ever taste Pennsylvania sweet corn that's been picked less than a half hour before you chomp into it? If not, you can't imagine it.

The local families weren't plain Amish. They were Mennonite, your fancy type of Pennsylvania Dutch. But that didn't mean they got confused about where their little John Deere tractors belonged on Sundays. They belonged at rest in the shed. Meanwhile, the farm families all went to church, and, after, had a proper Sunday dinner.

We did have telephones. Party lines shared by maybe eight, maybe ten families. It was easy to catch up on what was going on. Except for Korea, not much was.

Most folks spoke Pennsylvania Dutch on the phone. I couldn't. But my grandma's favorite language was Yiddish so she could easily understand the Dutch. I suppose they could have understood her too if she'd only left out all the Hebrew words and spoken without her thick Lithuanian accent. But she never did know the difference between the Hebrew and the German words. It was all just Yiddish to her. As for her Lithuanian accent? Hopeless! Sometime in the early '50s, she ate her first slice of pizza pie. She looked disappointed. Carefully mispronouncing the name of that fruit that is the pride of Georgia, Grandma complained, "Where were the pitzas?"

My grandpa, a butcher, was usually stuck back at home in the hot city on weekends. But on lucky Fridays, he would take the interurban from Reading Terminal 35 miles to Souderton. We'd drive the two miles from Harleysville and pick him up. He spoke four languages but he didn't drive.

The interurbans were just heavy trolley cars. I could never understand why all the adults called them trains. Trains are supposed to be big and black, puffing smoke, hissing steam, and snorting like the bull in the pasture across the fence. That's what you saw on the Lehigh Valley RR hauling all that coal past us during the war. I wanted to drive a steam locomotive. Still do, although I've given up on the idea of doing it for a living. I know that coal is pretty dirty and the war was the worst thing that people ever did to themselves—so far. But I loved those trains.

Mom did drive. She learned in 1949. But Dad, a G.P., was generally off earning our keep with the family car. He was making house calls. So if we kids wanted to get to town, it was shank's mare for a mile.

We liked to go to Henning's. That was Harleysville's general store. It was a dark place, not lit up with bright fluorescent lights like the A&P supermarket. It didn't matter. We really didn't need to see anything in Henning's. Mr. Henning was the only one allowed into the stock. You asked, and he got it for you. If he had it. And he usually did. He knew all the prices by heart. I bought my first chess set in Henning's.

Some things you didn't get at Henning's. A fellow in a van brought them right to your door. Bread for instance. Ice came that way, too, if you needed it. We had an electric refrigerator, though. And milk came by van, not only in the country, but even in the city where we lived during the rest of the year. The horse that pulled the milk wagon knew the route better than the milkman. Abbott's milk wagons were the last thing to get motors in Philadelphia.

Toward the end of the summer, Dad would start driving us to Heckler's place. That's where the farmers sold their produce. And that's where you could get the freshest vegetables and the sweetest watermelons. It wasn't a big place. Didn't need to be. Five miles down the road in Green Lane there was another store pretty much like it. Holzappel's, I think it was called. All the little towns had a little store for farmers to sell their fruit and vegetables. The only big place was the Quakertown farmer's market, about 15 miles away. The farmers used it to sell direct from little stalls. You never tasted better apple cider anywhere.

The farms aren't there anymore. They were done in by a combination of good and bad things that happened to us all. Mostly, I guess, middle-class people got to expecting more. More cars. A place in the country. Private telephone lines. Even cell phones. They got used to daily commutes longer than our annual trip to Harleysville. They put up with managed-care health plans and hospital emergency rooms. They eat fruits and vegetables that look as good as wax stage props—and taste about the same.

Farming changed, too. Mostly, it got bigger and more complicated. (For all the changes, I haven't noticed that the watermelons taste any sweeter.) Anyway, the county had to raise the farmers' taxes because property values went up sky high. The farmers got so rich they had to sell out. Land rich and cash poor. I guess they didn't really care to sell out, but they had no choice. Their farms are now subdivisions. Wall-to-wall houses.

Maybe old-fashioned farm life wouldn't be worth much to us folks now. Maybe whoever misses it is just showing his age. But it's easy for people not to miss what they never had and never saw. When we lived in the city, our street of duplexes was lined with sycamore trees. And across the street was a nice estate with lots of different birds and shrubs. Now all the sycamores are dead and gone, killed by dirt in the air. And Philadelphia paved over the estate for a school yard. I guess nobody who lives on that street today misses those trees and those birds. But I would.

How much is it worth? Who knows? Nice folks in a nice place to live? Tasty corn on the cob in August and a few blue jays across the street? Who knows what they're truly worth?

CHAPTER **6**

Hard-Core Reconciliation

> Progress . . . depends on the
> encouragement of variety.
> Calvin Coolidge[1]

Carole and I were sitting in the shade of the second floor gallery of the Casa de las Conchas in Salamanca, Spain. Besides the geometrical elegance of the gallery itself, all we could see were the red-tiled roofs that typify so much of the country. Suddenly a small, dark, and handsome thrush hopped onto the stone railing across the courtyard. Its flashes of auburn told me it was a male black redstart. The bird quietly went about its foraging business in full view of us tourists, evidently paying no attention to anything but bugs and perches. I could not see whether any insects were there for the taking, but the bird was acting as if its belly were being loaded. Soon it flew away.

Amid all that old tile and stone, here was a lovely, wild thrush. I have seen black redstarts on the ocean's shore. I have seen them in the desert's

A black redstart sings from his perch atop a ceramic roof tile in a French garden. Digiscope image by Alain Fossé.

stony solitude. I have seen them even among the high-mountain, rubbly wilderness above the tree line. And here they were coexisting with civilization in a most unpromising habitat—touristic monuments in a city center that had not seen its natural environmental state for perhaps 500 years.

Some thrushes are like that. The European blackbirds and American robins strut about fearlessly and successfully in almost every patch of grass we plant for ourselves. But some thrushes—like Townsend's solitaire—are quite shy of people, hardly ever seen except by the intrepid sort that seek out the Earth's most pristine habitats.

Many birds find homes in our habitats. For instance, Little and Crowe counted birds in and around the apple orchards of South Africa's Elgin district.[2] The apple orchards—especially those that use green methods of pest control, such as natural enemies—do not turn out to be so bad for birds: 7 species lived in the heavily sprayed orchards, 12 in the green-managed orchards, and 22 in the patches of natural vegetation left around the orchard landscape. A total of 24 species lived in at least one of these three situations.

But Little and Crowe observed 30 bird species in their work, not 24. Where did they see the other six species? They saw them in a nature reserve not far from the orchards. The reserve lies in the heart of the *fynbos*, a magical heathland habitat of immense plant diversity. Fynbos gives us a very large number of flowering houseplants and flowerbox species, such as geraniums. Until European settlement, all of the Elgin district was fynbos.

Twenty-four of thirty species. So, converting fynbos to apple orchards had wiped out six of the thirty species. In the Elgin district, they now live only in the fynbos reserve. Two of these—the orangebreasted sunbird and Victorin's warbler—are strict specialists. When enough fynbos goes, the world will lose them altogether.

German has words to distinguish between animals that we help—or at least that live with us successfully—and those that avoid us like the plague that we are. Species that endure or profit from the habitat changes we wreak on the face of the Earth are called *kulturfolger*, "culture followers." Those which will have nothing to do with us are called *kulturmeider*, "culture avoiders."[3] Those six special fynbos species are *kulturmeider*.

Words may help us communicate. But they can also—if we let them—freeze our perceptions of the world. In this case in particular, they can fossilize our attitudes to species. *Kulturfolger*—these we help, and we do it simply by being ourselves. No need for conservation measures. *Kulturmeider*—these we hurt, but we cannot do much about it. It is part of their nature. God has popped them into the box of species that avoid us. Too bad. Conservation measures won't help either, except perhaps in reserves.

Nonsense! Reconciliation ecology proposes to make *kulturfolger*—culture-following species—out of *kulturmeider* species. I know we can do it, at least for some species, because we already have. The distinction between *kulturfolger* and *kulturmeider* species is not so rigid after all. In this chapter, I will show you some examples.

Nest Boxes for Bluebirds

We have already considered the thrush family. We labeled as a *kulturfolger* some thrushes like American robins. In contrast, Townsend's solitaire is a *kulturmeider*. But it's not always so clear where a species belongs. Consider the case of the eastern bluebird.

Bluebirds are thrushes, too. They hunt insects in open fields and shun deep, primeval forests. They like nothing better than the bucolic scenery of rolling farmland. At one time, they thrived in such human habitats. They nested and fledged their babies successfully in school grounds, in front lawns, and along country roads. *Kulturfolger* without a doubt. Or so it would seem.

But when was the last time you saw an eastern bluebird? In fact, very few of them remain, and most of those live far from people. Once, it was one of our commonest birds. It nested in residential areas of cities and towns. It covered a vast range from Quebec and Nova Scotia in the northeast, west to the plains of eastern Montana, Wyoming, and Colorado, and south throughout the USA from Florida to Texas. What went wrong for the bluebirds?

Some may trace the problem to a severe winter in the southeastern U.S. in 1957–58. Bluebirds suffered terribly. Indeed, their populations never recovered. But 1957–58 was not the first bad winter for bluebirds. Going back only a bit more than a century, we find such winters in 1894–95, in 1939–40, and again in 1950–51. After each previous debacle, the bluebirds—prolific breeders—recovered in only a few years.

Part of the reason for bluebird decline is that bluebirds nest in holes. In a primitive world, dead and dying trees provided these holes in abundance. But dead and dying trees offend our sense of order and beauty. So, we manicure our properties and deprive hole-nesters of homes. Moreover, for several human generations, some bluebirds found replacement holes in our wooden fence posts. But now we use steel posts and they are totally useless to bluebirds (and a number of other species).

But shortage of holes for nests is just the beginning of what ails bluebirds. Their real malady is two aggressive, alien, outrageously abundant bird species that also like to nest in holes. House sparrows and starlings.

Did you know that we deliberately brought the house sparrow to our shores? It seems hard to believe now, but we did. Starting in 1851, we transported and released a series of shipments of the pests in various places along the eastern seaboard. On purpose! They thrived and, at first, bird enthusiasts delighted in them. But soon the threat they posed to native species ruined any pleasure we might have taken in them. Consider the contemporary witness of E. Howard Eaton:

> In the year 1879 there were no English [*i.e., house*] sparrows in the village of Springville (NY), where the author's boyhood was spent. That winter he visited the city of Buffalo and was delighted to see the English sparrows about the streets and dooryards amid the deep snow in the coldest weather. Two years later the sparrows had thoroughly established themselves at Springville and before the year 1888 had occupied practically every hamlet in the state. During the last 20 years they have been working their way from the cities and villages into the country and nearly every large farmyard is thickly inhabited by these troublesome parasites.
>
> ... the sparrow builds so early in the season that nearly every available box and hollow limb is occupied by the time the bluebirds, chickadees and nuthatches, martins and Tree swallows begin to think of their nestbuilding, so that the scarcity of nesting sites, which becomes greater and greater in all civilized communities, is multiplied tenfold by the occupation of all the available hollows by the indefatigable sparrow.... Thus the number of bluebirds and martins that nest in our dooryards or about the village is becoming smaller and smaller.[4]

One might have thought that house sparrows were bad enough. But in 1890, an ignorant do-gooder named Eugene Schieffelin brought a small group of European starlings to New York's Central Park and released them. The country will never be the same. Eaton (see p. 220) did not dream that starlings "could ever become the pest that the English sparrow has proved itself in all parts of the country." But they have. Starling populations have exploded and they live just about everywhere that people do.

To bluebirds, starlings are a pox even worse than sparrows. Like sparrows, starlings also nest in holes. But starlings are aggressive and larger than bluebirds. Starlings evict bluebirds from any nest site they want for themselves. As if that were not bad enough, starlings eat the berries that the bluebirds need to survive during the winter.

Eastern bluebird and a bluebird nest box. Drawing by Jack R. Schroeder.

Eaton feared the result would be a bluebird disaster and he was right. Any bluebird nest in the vicinity of agriculture or human habitation was doomed to pass to sparrows or starlings. The bluebird population plummeted. In a few decades, it had disappeared almost entirely from our company. Sparrows and starlings had changed an abundant *kulturfolger* to a scarce *kulturmeider*. What a disastrous step backward!

But devotion, diligence, and reconciliation saved the bluebird.[5] First, people discovered that a nest box with a hole 1½ inches in diameter suited bluebirds but excluded starlings. Then they found that house sparrows do not like shallow boxes; a nest box only four or five inches deep discourages them but not bluebirds. A compendium of sparrow-thwarting, starling-foiling nest-box designs are now available.[6] In 1979, the North American Bluebird Society was founded to spread the word and to encourage people to deploy appropriate nest boxes on their property. Bluebird numbers began to build back.

Today's North American Bluebird Society has expanded its goals. It still promotes the recovery of the Eastern bluebird, but it has spread its wings to cover both Western and Mountain bluebird species, as well as other bird species that need cavities for nest sites. It sponsors research into better nest box design. If you log on to its website,[7] you will see a variety of nest box plans and construction details.

The society publishes a magazine, *Bluebird*, that four times a year disseminates the latest information about supporting hole-nesters. It educates the public about ways to limit the damage that comes from house sparrows. And it explains the importance of keeping house cats under control and away from nesting bluebirds.

The National Wildlife Federation also lists the specifications for the nest boxes of many other bird species. These can all be placed in appropriate areas to supply a key requirement for the successful reproduction of wild birds. Moreover, in many cases, research proves that the boxes do not just attract birds, they actually increase their populations. We've already seen that holes boosted Eglin Air Force Base's population of red-cockaded woodpeckers. Nest boxes added to swamplands did the same for America's wood ducks, too. And various studies in Europe have identified a number of other species—like pied flycatchers—whose populations rise dramatically after nest boxes are installed.

Not all birds nest in holes, but those that do cannot find enough of them. People usually do not tolerate the dead or diseased timber that used to furnish these holes in the days before chain saws. But they do not mind the nest boxes that we can use to replace these scarce resources in a reconciled habitat. What delight they find when a bluebird or a purple martin family decides to take up residence in a nest box set out on their own land!

No doubt few species are attractive enough to inspire the sustained devotion of a large group like the Bluebird Society. But luckily, most species will not need such a single-minded effort. Yes, they will need some of us to adjust our habitats to make room for them to flourish. But unlike bluebirds, they will find their way with the help of less structured organizations, like neighborhood gardening alliances. Or they will find homes in government-sponsored reconciliation efforts like the one at Eglin Air Force Base. But they will succeed if we help them.

Perches for Butchers

Possibly you've never even heard of shrikes, but they are remarkable songbirds. They are often called butcherbirds. Why? Because of the way male shrikes advertise their prowess during the breeding season. They impale the corpses of their prey on twigs, thorns or even the barbs of an old-fashioned barbed wire fence. Potential mates get turned on by these grisly displays. Did I tell you about the black feather masks that shrikes wear over their eyes? I am not making this up.

Butcherbird displays give you a fine idea of a shrike's varied diet: mostly larger insects like grasshoppers, dragonflies and bees, but also spiders, lizards and even the occasional mouse. If you're in the right size range for a shrike to hunt, look out!

So, if they're so good at eating all sorts of common things, why aren't they all over the place like sparrows and robins? Because shrikes prefer to hunt in a certain restricted way. They like to sit up on a post or branch and scan the ground around them. Then they pounce.

A loggerhead shrike.
Drawing by
George Miksch Sutton.

Compared to birds that do hunt on the wing, shrikes are not spectacular fliers. So they rarely take prey that is airborne—unless it is something slow like a butterfly. They also rarely hover to scan the ground below. Hovering is very hard work, and usually makes their meal too expensive. Would you run a marathon for a slice of bread?

Because of their hunting style, shrikes need a semi-open landscape. Low vegetation, like grasses, interspersed with shrubs from which to scan and pounce. Not a terribly scarce landscape at one time, and it proved a recipe for success in both the Old World and the New.

The world has 30 species of shrikes. In the Old World, they range from Japan to Norway and down to South Africa. Two species live in North America. Australia has no true shrikes—the very similar looking Australian butcherbirds are unrelated. South America also lacks shrikes.

Today we must face the very real possibility that all species of shrikes are on the road to extinction.[8] In European and American countries where bird counts are a tradition, shrike populations have declined by over 50 percent. Some countries have lost entire species; Switzerland, for example, has lost two of its four. So has the Czech Republic.

Most likely, a variety of changes precipitated the decline of the shrikes. The soft edges of small farm fields, with their hedgerows and scattered trees, gave way to the sharp edges of modern agriculture: corn to woods in two meters. Perhaps, also, pesticides poisoned the grasshoppers, crickets and beetles that form the bulk of shrike diets. And, whereas the barbed wire fences of yesterday were supported by wooden fenceposts that made great shrike perches, today's fences use thin steel posts, which are not so good for a shrike's business ventures.

Whatever the underlying causes, there is all too real a chance that our great-grandchildren will not be able to see any shrike—except in books alongside the pictures of other once-common marvels that we have swept aside. But there are many people who have decided never to let that happen.

Among the most active is my friend Reuven Yosef—you remember him from the pseudo-salt marsh he built in Eilat. Yosef was the guy who suggested that male butcherbirds impale their prey to show off for their mates, and then he backed up his interpretation with convincing scientific evidence. Believe it or not, this common and ostentatious behavior had gone without its true explanation since the dawn of natural history. Reuven nailed it.

Yosef did his Ph.D. on a very special working ranch in the middle of southern Florida. The MacArthur AgroEcology Research Center is a unit of Archbold Biological Station. It's not called a ranch for old-time's sake; its 10,000 acres *really* are devoted to cattle raising. It has a bunch of cowboys, a ranch house, horses, pastures, corrals. I've seen the place myself, and I can tell you that—except for the high humidity—it makes me feel right at home, almost as if I were in Arizona.

Yet, as you surely guessed from its name, the MacArthur AgroEcology Research Center is not just a ranch. Its job goes beyond beef. The folks at Archbold want to discover how to make money from cattle that graze in an ecologically sound landscape. Keeping diversity high is, for sure, a part of that. Sounds like reconciliation ecology, doesn't it?

Yosef studied the ranch's population of loggerhead shrikes.[9] That's the more southern of North America's species, and it is certainly one of those in trouble. He banded them, weighed them, watched them hunt. He studied the times they nest and the number of times they would try to re-nest if weather or a predator destroyed their first effort in a season. He followed their babies' progress as the parents struggled to bring up their own. Experienced shrikes seemed to accept him as a natural phenomenon—if not a member of the family. I have seen him call them, and watched them come promptly to his field vehicle for a treat of mouse flesh.

From his work, Yosef figured that a lot more shrikes could use the pastures. There were plenty of big insects where no shrikes lived, but the pastures had very few perches. He went to a lumberyard and got some fence posts. He stapled a little bit of barbed wire to one end to help the shrikes keep their footing. Then he installed them in a pasture where shrikes lived but did not use much of their territory for hunting. Being a card-carrying scientist, he did not do this to all the shrike territories, but only to half, keeping the other half as the controls. The most difficult part of his job was convincing the cowboys not to climb in their jeeps and use the fence posts as a slalom course.

Here was the idea.

Yosef guessed that those shrike territories were as holey as a Swiss cheese. The birds were hunting only near a proper perch—that was the cheese part—and ignoring the parts too far from a perch—that was the holey part. He wanted to use the fence posts to fill in the holes.

If he succeeded, the shrikes should shrink their territories. Why? Because territory is expensive to defend. Not only that. It also costs a lot to fly from your nest to your hunting perch and back, especially when the perch is far from your nest. So why defend a territory larger than you need?

And if the shrikes did shrink their territories, that would make room for more shrikes. Aha! Yosef was really conducting an experiment to bolster the population of a bird in trouble.

Results came quickly. Within the first spring, territories with extra fence posts shrank dramatically. On average, the experimental territories were 77 percent smaller. The minimum shrinkage was 68.6 percent, and one had shriveled up by 83.9 percent. But the controls had stayed the same size.

As expected, new shrike "homesteaders" settled in the land left vacant by the smaller territories. The loggerhead shrike population increased 60 percent.

The smaller territories brought another advantage to the nestlings. They helped them survive. Parent birds in smaller territories had 33 percent more successful clutches than controls, and raised 29 percent more chicks per successful clutch. Clearly, oversize territories do hurt.

It all amounts to a recipe for reconciliation. Now we know how to raise the reproductive success of loggerhead shrikes and provide them lots more habitat in a working cattle ranch. There is no need for these shrikes to join the list of the doomed in the *Red Book of Rare and Endangered Species*.[10]

Many other species of shrikes in trouble are also being helped. Yosef's methods seem to work for the fiscal shrike of South Africa.[11] In Belgium, Dries van Nieuwenhuyse has increased the population of the redbacked shrike with a technique quite similar to Yosef's, working in a hilly area used for cattle breeding. He trims shrubs and hedges, piles up dead branches, and even sets out balls of barbed wire in strategic locations. But he does not ask the cattle to go away![12]

In Germany, Martin Schön has performed similar miracles for great grey shrikes.[13] He alters the monotonous agricultural landscape and makes it a patchwork of microhabitats, including the great piles of stones that farmers used to leave at the edges of their fields as they plowed them up. Shrikes soon return to breed successfully where they have not bred for many years.

Shrike reconciliation ecology teaches us some general lessons. First, drink deeply from the natural history of the species you want to help. Study their reproductive cycles, their diets, and their behavior. Abstract the essence of their needs from what you observe. Then apply it without worrying whether your redesign of the human landscape will resemble a wilderness. It won't, so feel free to be outrageously creative. Birds and other animals appreciate abstract art more than you think!

Natterjack Toads

Reconciliation ecology is not always going to be as easy as a taking quick trip to the lumberyard. Ecologists may need to do a lot of research, perform a lot of trials, experience a lot of error. Consider the case of *Bufo calamita*, the natterjack toad.

Natterjack toads are rare and endangered in the United Kingdom. To preserve the species, 50 ecologists engaged in a detailed, multifaceted and sustained effort over the course of a quarter of a century. Their studies of the natterjack toad constitute a model of what may often need to be done and how remarkably successful we can be in doing it. Those studies have culminated in the development and installation of habitats that are saving the natterjack toad in the United Kingdom.[14]

The work of this team began by finding out what natterjacks do for a living. Yes, of course they eat insects, but in what special habitat? The natterjack toad turns

Bufo calamita, anonymous.

out not to do well in thick, tall vegetation such as birch, gorse, and bracken. Instead, it pioneers and thrives in more open plant cover such as that surrounding the rich pools of coastal dunes and the poorer ones of inland heaths.

Natterjacks have a fierce competitor, *Bufo bufo*, the common toad. But many things distinguish it from its competitor and permit it to stay in business. Unlike common toads, it burrows in sand. When foraging at night, it operates at a body temperature 1.4 degrees Celsius higher than *B. bufo*. Such higher temperatures prove crucial to natterjacks. They lose weight if forced to forage in dense, cooler vegetation. This helps to explain why its population declines if tall vegetation begins to invade and shade natterjack habitat. The increased shade also lowers the water temperature of the pools, slowing the development of natterjack tadpoles and subjecting them to more damaging competition from *B. bufo*.

In developing their picture of good habitat, the research team studied a very long list of environmental factors. They looked at a single-celled parasite, *Prototheca richardsi*, that lives in natterjack guts. They measured predation by salamanders, Odonata, water beetles, water bugs, and Notonecta larvae. They also studied pond chemistry and water quality (chlorides, sulfates, orthophosphates, ammonia, iron, sodium, potassium, calcium, magnesium, alkalinity, conductivity, color, and turbidity). They even studied pond depth and the contour of pond slopes.

Soon they began to advance schemes to redesign part of the dune world on behalf of natterjacks. They cleared dense vegetation and reintroduced grazing to prevent its regrowth. They fought acidification of the water by adding slaked lime (calcium hydroxide) to natterjack ponds every year or two, or by scraping most of the sulfate-rich silt from the pond bottoms. To give the natterjacks a fair start, they even removed some common toads.

Most amazing to me was that they had learned enough to build successful natterjack ponds where none had been before. They used old bomb craters and active golf courses, and they built some 200 new ponds—not too deep, for that would have encouraged invertebrate predation—and not too steep, for they need a lot of shallow, warm water—and sometimes lined with concrete to fight acidification.

At all sites with new ponds, natterjacks used at least one and usually most within a year or two of construction. The new ponds rescued or increased natterjack populations in two-thirds of the sites. New research

©Anna Flatten '84

This bat roost near Comfort, Texas, was built to increase the local bat population in order to fight malaria-carrying mosquitoes. About 1,000 bats of two species currently live in it. The roost was designed by Charles Campbell and built in 1918 by Albert Steves, former mayor of San Antonio. During Steves's time as mayor, San Antonio resolved to protect its bats and made funds available for a municipal bat roost in San Antonio. Once, 16 bat roosts existed in the U.S., but today only this one (a National Historical Landmark) and another, on Sugarloaf Key, Florida, remain. Not far from Comfort, Bat Conservation International offers help and plans for bat roosts on a smaller scale. Drawing by Anna Flatten © 1984.

results are explaining the failures and supporting retries but nearly everything about the story is encouraging, including the great pride that local residents have taken in the work. Even the grazing has turned a small profit!

Cattle Tanks for Chiricahua Leopard Frogs

Frogs and toads are in the news a lot these days. Unfortunately, the news is usually bad. Like the shrikes, many of their species are in danger of extinction. Scientists are not sure why, but in one case we know exactly what is going on. We even know what to do about it. It is the case of the Chiricahua leopard frog.

Chiricahua leopard frogs are quite ordinary four-inch, green and gold frogs that live in remote corners of southeastern Arizona, southwestern New Mexico, and northern Mexico. Once, nobody paid them much attention. In their scarce local habitat—streams running through desert canyons—they croaked away in abundance. For the last few decades, however, they have just croaked. They have disappeared from about 75 percent of the places in which they lived only a short while ago. On 13 June 2002 (perhaps auspiciously, a Thursday rather than a Friday), their plight was officially recognized—the government of the United States declared them "threatened."

The trouble began with their small size. Compared to bullfrogs, they make slim frog's legs. So, Man brought in the bullfrogs and set them loose. Once we introduced them, the bullfrogs spread rapidly with no further help.

Disaster. Bullfrogs eat other frogs. They eat their own species and other frog species. And they eat a lot of them. In one study, conducted by Phil Rosen and Cecil Schwalbe on the San Bernardino National Wildlife Refuge (Cochise County, Arizona), about one-third of the bullfrog's diet was other frogs![15] So, you should not be surprised to hear that Chiricahua leopard frogs vanished from the San Bernardino National Wildlife Refuge by the late 1980s.

Bullfrogs are not the only introduced enemy destroying leopard frogs, shrinking their range and threatening their prospects for survival. Green sunfish, with their appetite for tadpoles, cause problems, too. But the leopard frogs did have one last chance. Neither bullfrogs nor sunfish inhabit the isolated stock tanks that ranchers use to water their cattle. In addition to frogs and cattle, many other species of wildlife use these tanks.

The Magoffin Ranch lies a few miles east of the San Bernardino National Wildlife Refuge.[16] In the early 1990s, it still had two populations of leopard frogs. They were living in water holes on the ranch. One, called Rosewood Tank, almost lost its leopard frogs during severe droughts in both 1989 and 1994. In fact, without the cattle, it would have. Both in 1989 and early 1994, the watery mud that accumulated in the cattle's hoof prints actually saved the leopard frogs from death. But the 1994 drought got even worse during the spring. By April, the leopard frogs were doomed. Except, the Magoffins refused to quit.

They dug an unnatural ten-foot-deep pool in the bottom of Rosewood Tank as it dried. Then they collected 400 leopard frog tadpoles from the drying tank bottom and moved them into their new hole. Every week throughout the dry summer, they hauled water to the hole in a 1,000-gallon tank truck and kept the frogs alive.

Owner Matt Magoffin has since dug out the bottom of the main tank so that it holds more water. And he has changed the outflow channels to improve the tank's depth. When the next drought hits, Rosewood Tank should be about 12 to 15 feet deep. Meanwhile, the leopard frogs are doing well. And so are the cattle.

Once he saw what he could do, the Magoffins expanded their efforts at reconciliation. They tackled a second water hole on the ranch. Belency Tank had dried completely during the drought of 1989. The frogs recolonized naturally, but the Magoffins saw that such flirtations with extinction were bound to lead to a calamity in the future. So, they dug a well for Belency Tank to prevent the recurrence of a total frog kill. Belency Tank now has about as many leopard frogs as Rosewood. The Magoffins then renewed a smaller tank, Choate Tank, with a windmill to pump water and a small pond, thus producing the support required for a third population of Chiricahua leopard frogs on the ranch. Efforts to reintroduce the frog to the National Wildlife Refuge depend on the Magoffin populations for colonists.

The Magoffin Ranch has joined with several other ranches along the Arizona-New Mexico border just north of Old Mexico. The association is named the Malpai Borderlands Group.

The U.S. Fish & Wildlife Service termed the ranchers' role in frog recovery "crucial." Moreover, in its listing of the frog as threatened, it specifically exempts livestock tanks from the usual prohibitions associated with a threatened species.[17] Approximately half the Chiricahua leopard frogs alive today live in stock tanks or artificial reservoirs. Thus, the listing of this species appears to represent the very first time that a national government has recognized the importance of reconciliation ecology in ministering to an endangered species. Hopefully, there will be many more such acknowledgments.

In truth, the Malpai Borderlands Group is much more than a society for saving frogs. They are out to preserve ranching as a way of life. But in doing so, they have quickly come to realize that they must preserve the land on which that life depends. They must save its species—frogs and snakes, pronghorn antelopes and native grasses. They must manage its fires, care for its soil, cherish its water.[18]

One day, the light dawned, and they saw that they were conservationists.

In the west, most of the time, ranchers and conservationists tend to tear at each other's throats. A rancher-conservationist? What a contradiction in terms! The Borderlands ranchers must have felt like they had split personalities. Prognosis? Probably being on everyone's enemies list.

Undaunted, they got together with some flesh and blood conservationists. They shared a concern that the grazing lands of the west be preserved in ecological health. Soon, the ranchers learned about conservation easements, and the conservationists learned that nobody was taking better care of their environment than those ranchers. What a surprise! Everyone got along real well.

In the rangelands of the west, conservationists reflexively oppose ranchers as environmental enemies. And ranchers see environmentalists as demons trying to deprive them of their land and way of life. The Malpai Borderlands Group has shown us that reconciliation ecology can change all that. The Malpai Borderlands Group has established the "radical center."

CHAPTER 7

Happy Accidents

> Amazing! I've been speaking prose
> for more than forty years
> without ever knowing it.
>
> Molière[1]

About 2,100 years ago, Rabbi Simon ben Shetah was in the market for an ass. After some bargaining with its Arab owner—and probably some refreshment—he brought his new beast home, complete with its harness. When he arrived, his students helped him with it. Suddenly, they shouted, "Rabbi, 𝕲𝖔𝖉'𝖘 𝖇𝖑𝖊𝖘𝖘𝖎𝖓𝖌 𝖊𝖓𝖗𝖎𝖈𝖍𝖊𝖘."[2] Suspended around the animal's neck, they had discovered a precious gem.[3]

Sometimes you get more than you pay for.

But Rabbi Simon was a member of a wealthy family, the brother of Queen Salome Alexandra of Judaea. Perhaps that and his renowned piety help explain the rest of the story: Declaring, "I bought the ass and not the gem," Simon returned the gem to its owner.

Of course, Simon wanted to teach honesty and kindness. He knew that keeping the gem would likely have done great harm to its rightful owner. But sometimes that is not true. Sometimes, you get more than you pay for and you harm no one if you keep it. In some cases of reconciliation ecology, Simon would have decided that you must keep the precious bonus. Giving it back would do a lot of harm. These are the happy accidents.

American Crocodiles in Turkey Point Power Plant

The Mesozoic era (about 248 million years to 65 million years ago) is known as the Age of Reptiles. The ruling reptiles, or archosaurs, that dominated the Earth during much of the Mesozoic include the world's favorite fossils, the dinosaurs. But other archosaurs also flourished. Two of these have left descendants in our time: birds and crocodiles. They lie closest to the dinosaurs on the tree of life.

Actually, three quite different sorts of crocodiles survive. The gavials have long narrow snouts and specialize in eating fish. The alligators have very broad snouts, so powerful that they can crush and eat turtles. The crocodiles with intermediate snouts are called simply crocodiles. You can recognize them easily because their snouts aren't quite broad enough to hide their teeth. Thus, their fabled winning smile. By comparison, alligators seem dour and closed-mouthed.

The United States has no gavials, but it does have alligators aplenty. It has a few crocodiles, too, members of a rare and endangered species that ranges from the United States into the Caribbean and down through Central America to tropical South America. All the U.S. crocodiles live in tropical Florida where they are very rare indeed.

I have seen the Nile crocodile in Africa and the huge saltwater crocodile in the Australian outback. But I have never been lucky enough to spot an American crocodile. On one trip to Florida, I found out why. I had been looking in the wrong places. I should have tried the interceptor ditch of the Turkey Point power plant south of Miami.

The Turkey Point power plant, operated by the Florida Power & Light Company, has the job of generating electricity for southern Florida. It consists of two fossil fuel generating units and two that run on nuclear fuel. Turkey Point emits a lot of hot water.

To cool off its effluent, Turkey Point dug an extensive system of canals. The system covers 6,000 acres, and about 64 percent is open water. The shallow canals, each approximately 200 feet wide, run through the landscape like densely drawn zebra stripes from north to south. If placed end-to-end, they would be some 160 miles long.

To separate the 38 canals, the plant piled the dirt dredged to dig the canals into low-lying berms about 80 feet wide. The berms support a variety of native and exotic plants, including buttonwood trees, red

Turkey Point power plant, southern Florida. The plant is half fossil fuel and half nuclear. It should be obvious that this place was built to mean business—to generate electricity without compromise! Photo courtesy Florida Power & Light Company.

mangroves, and casuarinas. Red mangroves also grow along the edges of the canals themselves.

We can all agree that the canals of Turkey Point were built to cool water, not to support diversity. If you had asked me, I would have guessed that they would be a terribly poor prospect for conservation. I would have been sadly mistaken.

In addition to the variety of plants growing on the berms, a large, healthy population of American crocodiles lives in the cooling canals. No one planned it. It just happened.

Although it did not intend to get into the conservation business, Florida Power & Light Company is behaving like a model citizen. They employ biologists to monitor the crocodiles and do what they can to ensure their continued success. The biologists discovered that their crocs reproduce well. In fact, they are responsible for producing about 10 percent of all new young American crocodiles in the United States.[4]

Cooling canals at Turkey Point. Crocodiles do well in the warm water and sun on the berms separating the canals. Photo courtesy Florida Power & Light Company.

Perhaps this story has a moral. Never give up on a new habitat, no matter how dim its potential would seem to be. Life is remarkably tenacious and opportunistic. A little change here, a little addition there, and who knows? The habitat may get just what it takes to save a rare and marvelous species.

And even if that is the wrong moral, no one would suggest that the crocodiles interfere with the physical work of cooling the water. The cooling canals are reconciled with the needs of a very cool reptile. We should accept the accident of their reconciliation with a smile as broad as that of the crocodile itself.

Like a Las Vegas gambling casino, reconciliation ecology often depends on happy accidents. The Earth has at least two million—if not 20 million—species to save. We haven't the time, the space or the money to do a habitat study and reconciliation project for all of them. But if we stack the deck by improving the habitat for some of the species, we are

88

going to win the continued existence of many others. The odds are that what one species finds tolerable, others, especially those once found with it in primeval landscapes, will also manage to tolerate. Consider the Bargerveen.

The Bargerveen

The Bargerveen is a relic of an ancient ecosystem, the peat-moor of Europe. It is a small, remnant patch of a huge peat-moor whose 300,000 hectares once sat astride the Netherlands-Germany border. Today, all but the Bargerveen's 2,000 hectares have been turned into farmland. The Bargerveen, situated entirely in Drente Province, in the Netherlands, has been set aside as a reserve.

Peat-moor supports many plant species. At least it did two centuries ago. But, by itself, the Bargerveen cannot. It is simply too small. Its lack of soil moisture makes the problem obvious. Peat-moors are meant to be very wet. Yet, thirty years ago, the Bargerveen's peat was rapidly drying out, its water draining off to the farms that surround it. As a reserve, the Bargerveen was doomed to fail.

In 1972, the Netherlands's Forestry Service began to build a 35-kilometer-long set of dams to hold the rainwater within the Bargerveen. That's a lot of dam for a little patch of peat-moor, but the Dutch resolved to save the Bargerveen, and halfway would not satisfy them.

Managing the Bargerveen has done the job. The peat-moor is coming back. Moreover, the management tactics include human use of the "reserve" too. In parts of the Bargerveen, people still dig peat for fuel. Peat mining produces areas that are considerably drier and support a very different set of plants compared to protected, wet peat-moor. As a result, many different habitats now dot the reserve. It has become a rich mosaic of semi-natural environments.

At least one happy accident has accompanied the management of the Bargerveen. Remember the plight of the world's shrike species? In the last chapter, we saw the success of reconciliation projects that targeted shrikes specifically. Now we will see a case in which, by accident, reconciliation has helped save a shrike species, one that lives in the Bargerveen: the last population of red-backed shrikes in the Netherlands.

Once common, the red-backed shrike has declined precipitously all over western Europe. By 1989, the last red-backed shrike had vanished utterly from Great Britain—believe me, if even one were left, the twitchers would have discovered it.

In the Bargerveen, matters were similarly grave. In 1981, only two pairs of red-back shrikes remained. Then something wonderful happened. Management strategies, designed to rehydrate the peat-moor for its plants, began to impact the birds. By 1984, the Bargerveen had 14 pairs of red-back shrikes. By 1992, it had 104! Incredible! Given a chance, nature's powers of recuperation can truly astonish us. Yet no one knows exactly why the shrikes of the Bargerveen have prospered.[5]

Hans Esselink and his colleagues, hard at work studying the shrikes, think the answer lies in the increasing variety, supply, and quality of shrike food. In that case, some of the insects of the Bargerveen must also have suffered a happy accident! Good. That is the idea. When the house wins, so do a lot of its occupants.

Growing Gefilte Fish

A thousand years ago, southern Bohemia was a pretty wild place. Rivers and forests? Yes. People? Hardly any. The region had a bad reputation. As long as 2,000 years ago, Roman Empire soldiers feared the primitive black forests. But today, they would hardly recognize the place. It is a bucolic farmland, dotted with villages, woodlands, and restful ponds.

People moved into southern Bohemia starting about 800 years ago. They transformed the landscape, clearing the forests and replacing them with their farm fields and woodlots. Then, without knowing it, they began a massive campaign of reconciliation ecology. We can see it most clearly by focusing on the Czech Republic's Třeboň Biosphere Reserve and Protected Landscape Area, situated on its border with Austria.[6] Třeboň BR covers about 700 square kilometers (270 square miles).

Třeboň BR got started in the mid-thirteenth century when the king of Bohemia granted a large tract of land to a noble family, the Vitkovec. They brought in settlers to help them clear and drain the land for farms. As the drainage progressed, the remaining water collected in ponds.

Undoubtedly, the ponds yielded some fish, and in the fourteenth century, during the reign of King Charles I of Luxembourg, the landowners decided to extend the ponds. They kept careful records so that we even know the exact year of some of these enlargements. In 1363, for instance, it was the turn of Dvořiště Pond to grow. By the beginning of the fifteenth century, there were 17 small ponds and 3 large ones covering about 700 hectares (1,730 acres). Then the reconciliation really started.

It all began with carp. Carp is the fish that today often ends up on the table as that Jewish soul food called gefilte fish. But Eastern Europeans in general love to eat carp. In 1450, family estate managers dumped some common carp of breeding size into one of the ponds. These bred and their larger offspring were harvested after six or seven years. They went to market and brought a good price. The family found itself in the fish business. They stocked carp in the other ponds, too.

> No fins, no flippers,
> the gefilte fish swims
> with some difficulty.
>
> — David M. Bader,
> *Haikus for Jews*

If 20 ponds could bring a good return, think what hundreds could do! The family ordered an intense pond-building program that was to last until 1584. They dammed the River Lužnice. They built a 46-kilometer-long canal to fill and drain ponds at will, and they even dug a new river. Today, Třeboň BR has 460 artificial fish ponds varying in size from less than a hectare to five square kilometers (1.9 square miles). They cover 10.6 percent of Třeboň BR's area.

By the end of the sixteenth century, the ponds were yielding 200 tons of carp each year. After some ups and downs in production, improvements in fish culture and the introduction of other fish species for harvest, the ponds passed to the hands of the communist regime of the twentieth century. Production rose dramatically. Today, with the ponds 90 percent in the hands of Třeboň Fisheries Ltd., a private company, annual production stands at an all-time high of about 3,000 metric tons.

But the Třeboň landscape does not look like a fish factory.[7] If you saw it, you would not know that the landscape is man-made unless someone told you. You would swear that it is one of the loveliest natural landscapes you've ever seen. As the centuries went by, nature worked hard on it and naturalized it. The area of the ponds and their surroundings became a semi-natural, albeit totally artificial, 173-square-kilometer wetland.

Reeds grow at the edges of its ponds and disguise their man-made origin. Canoeists and hunters take their pleasures from the beauty and the bounty of a landscape that nature and *Homo sapiens* wrought together. They immerse themselves in a world of 150 species of nesting birds, 50 species of mammals, 12 species of amphibians, and untold species of invertebrates and plants.

Třeboň BR earned its designation as a biosphere reserve on account of that diversity, not its success at producing fish. The diversity gets supported partly because the wetlands do not constitute just a single habitat, but an extraordinarily rich variety of them. Yet none of them would exist without the fish ponds and their infrastructure.

The high species diversity of Třeboň BR includes many species that are quite rare elsewhere. It is an important breeding site for butterflies and moths. Four species of its wading birds are rare or endangered, as are half its 16 species of bats. Some of its orchids and sundews face extinction elsewhere. It has rare cottongrasses and many rare ferns.

The treasures of Třeboň BR include white-tailed eagles (cousins of the North American bald eagle), European elk (cousins of our moose), great cormorants, and natterjack toads. Water violets grow in the canals. The list goes on and on.

And let's not forget about the otters. After all, otters are no more than large weasels that have turned into fish-eating machines. It will not surprise you to learn that a large area full of fish ponds also supports a lot of otters—about a hundred, in fact. Roughly 90 percent of the diet of the otters of Třeboň BR is fish.

Otters have disappeared from many European countries, and they have become quite rare and restricted in others. Moreover, otters are cute—drop-dead cute. Otters rank pretty high on the list of the dwindling charismatic megafauna. In other words, they are big-eyed, warm, soft, and in trouble. Added to all that, they love to play and we love to watch them. Lots of places want them back.

It sounds so perfect. An extensive, old, reconciled landscape in the heart of Europe. People farm it and live in it. They grow fish and timber in it. They make their living doing so. They also visit it for its natural features and delight themselves with its charm. Meanwhile it supports a large number of wild species, many of them rare, a few actually endangered. Perfect!

An otter, one of the world's most charismatic species. But it does eat fish! Anonymous.

Well, not exactly perfect. People raise the fish to sell, not to feed otters. Austria actually administers a program to reimburse its fish farmers for the depredations of otters. But the Czech Republic does not, which is probably why many Czech fish farmers wish their otters would move to Austria.

In fact, some fish farmers in Europe do more than wish. Although killing otters is against the law all over the continent, many otters are exterminated by their human competitors.

Fish farmers have long hated otters. "Otters are terrific, but not in my pond." From 1874 to 1896, people registered almost 1,400 otter kills in the area that is now the Czech Republic. Hunters and fish farmers set up an anti-otter booth at the 1891 national exhibition in Prague. Over it hung a grisly, crudely lettered sign: **SMRT VYDRAM** (Death to otters). The letters were fashioned from about 60 otter skulls and 20 otter lower jawbones. Clever. In Czechoslovakia, otter hunting was not banned until 1948.

The battle is not limited to otters. We have lots of competitors for the food we plan to harvest. Birds and insects eat our fruit as it hangs in the orchard. Small mammals and insects pilfer our grain. Raptors and carnivores make off with our richest sources of protein. Should we give any of these species aid and comfort?

Carnivores often seem to be the worst offenders. Sea otters grab the abalone more skillfully than the abalone fisherman. Sea lions and river otters take our fish. Wolves destroy calves and coyotes prey on lambs. Is there room for any of them in our world? Outside of zoos, that is.

Let's look closely at the case of the otters in Třeboň BR. You may agree with me that the fish farmers have overreacted. You will certainly agree that we have a lot more to learn about estimating the damage that otters do.

How much do the otters eat? Třeboň Fisheries Ltd., which controls some 90 percent of the fish farming in Třeboň BR, has the largest sample on which to base an estimate. They calculate that otters eat one million Czech korunas worth of fish from their ponds each and every year. That's about $32,500 (U.S.). Sounds terrible, right?

Actually, it isn't. First, it's only about $325 per otter per year. Most of us spend that much or more on our carnivore housepets. Second, otter losses amount to less than 1 percent of Třeboň Fisheries Ltd.'s annual sales. The great cormorants do twice as much damage! Yes, the number of cormorants are kept low—50 pairs—but no one seems to hate them, and no one seems to be calling for their exile.

Then how can we explain the clamor? Why are so many fish farmers calling for government payments to reimburse them for losses to otters? The outcry comes from the smaller fish farmer. Třeboň Fisheries Ltd. only once (winter 1995) complained about otter kills, and it turned out then that a disease, not otters, probably killed the fish.

But the smaller farmer heard about the Austrian payment system and about the reimbursements his nearby Austrian colleagues receive. Otters are surely eating his fish too. Fair is fair. Why shouldn't he get paid, too? And, since he hopes his demand will open a round of negotiations in which there is bound to be some hard bargaining, the Czech fish farmers blame otters for killing even more fish than the otters could possibly eat—and for always killing the largest, most expensive fish, as well.

Now get ready for a real curve ball. The otters may, in reality, be an economic boon to the fish farmer.

I do not deny that otters eat many economically valuable fish. But so far the accounting has been unjust. No one has bothered to give the otter credit for any good that it might be doing; the ledger shows only the debits. It's been all prosecution and no defense.

What defense? What credit? What good?

There are many different kinds of fishes in those ponds. Otters do not eat them indiscriminately. Despite the fact that they eat a lot of carp, otters prefer the other fish species. The reason may well be that carp contain thiaminase, which inactivates vitamin B-1. North American otters—very, very close cousins of the Eurasian otters—when fed diets very rich in carp, develop a life-threatening paralysis that appears to be the result of inadequate amounts of vitamin B-1.

Whatever the reason, otters take other fishes in higher proportion than carp. We can guess that these other fishes compete with the carp, or even prey on it. So, otters may do more harm to the carp's competitors and predators than to carp itself. On balance, otters might even enhance carp production.

A wild possibility? Not at all. Coyotes do have that effect on cattle. Rabbits severely reduce cattle forage, and coyotes concentrate on eating rabbits. Yes, they do eat a calf now and again—a visible behavior that has called down the wrath of ranchers on their heads. But the damage to calves pales in comparison to the indirect good that coyotes do by eating lots of rabbits.

Ecology bubbles with loony interactions like that, interactions that are full of surprises. One famous experiment concluded that a predatory starfish is actually necessary for the existence of the seven species that it eats most often. The starfish really likes to eat its prey's competitors. It likes them so much that it keeps them scarce. Without the starfish, the competitors of the prey would multiply profusely and obliterate them.

So, how much do Třeboň's otters cost? We do not know. No one has adequately measured the plusses and minuses. But the otter accounting problem is not as complicated as the soil-accounting problem set in chapter 5. And it is nowhere near as slippery as valuing the institution of the family farm. We can easily do experiments to answer the otter question.

Biologists know how to keep otters away from a pond: Put an otter-proof fence around it. They could do that to a few ponds of small to medium size. Then compare the fish harvests from fenced ponds with those of unfenced ponds. It is a straightforward and perfectly ordinary ecological experiment. How much fish do you get with and without otters?

What species can we protect with reconciliation ecology? Can we use it to protect any of those that compete with us? Perhaps we cannot manage to get along with all our competitors, but the state of our accounting leaves much room for improvement. An apparent competitor may turn out to be our economic friend. Or protecting it may turn out to cost so little and give us so much pleasure that we welcome it in our landscapes.

Bats in the Bauhaus

Walter Gropius founded the Bauhaus movement, one of the world's most severe schools of modern architecture. My father used to liken his designs to big chicken coops. Even if you don't agree with my father's opinion, you would have to agree that Gropius had little or no interest in imitating the natural world. Gropius was searching for a new and abstract reality for the human habitat. Stark, simple, geometric forms. Expanses of glass. White, texture-free cubes. Living and working spaces reduced to their ultimate functions. In the first third of the twentieth century, Gropius was all the rage.

In the 1920s, a number of young architects who had studied with Gropius found themselves together in the burgeoning city of Tel Aviv with plenty of commissions to fulfill. Instead of working independently, however, they colluded and produced White City.

White City is a unique neighborhood. Even today, it overflows with fine examples of Bauhaus architecture. Relieving the austere atmosphere of the neighborhood, a strip of green, meant mostly as a playground and promenade, runs through the middle of Rothschild Boulevard, White City's main artery. The strip also holds an incomplete sculpture garden—incomplete by design. The architects thought it would be meaningful if future generations could add to the garden and so express the artistic temperament of their time. And so they have.

To the north, Rothschild Boulevard ends in a monumental public culture center—Habima Square—containing the national theater, a modern art museum and a symphony hall. Its white buildings and geometric, paved courtyards fit perfectly with the artists' vision of a Bauhaus world. What little vegetation you see is planted—potted habitats for an artificial creation.

Yet, if you visit Habima Square on a summer's eve and look upward, you will not believe what you see. The air will be filled with Egyptian fruit bats!

I do not mean to mislead you. There are not so many and they are not so large as to make you fearful. But there are a lot of bats, and they are very, very large. Egyptian fruit bats are the somewhat smaller cousins of the huge flying foxes (with six-foot wingspans) that live in southeast Asia and Australia/New Guinea.

Most of the people bustling about the square do not seem aware of the fruit bats flying only five or ten feet above their heads. Nor do the bats seem to mind that they are flying in and out of the bright lights of the buildings in full view of the world's most untrustworthy species. Bats in the Bauhaus. I guess if a wild species can adapt itself to such an extreme human vision, there is hope for diversity yet.

Architects provided for the bats in Habima Square quite by accident, entirely without the deliberate intent of the builders of the hygeiostatic bat roosts pictured in the last chapter. Furthermore, one need not build in the style of Gropius to satisfy bats.[8] Many human structures give bats homes by accident. Attics and belfries. Barns and bridges. I have seen an established roost of Egyptian fruit bats in the tiny garden of the old downtown campus of Ben Gurion University in Beersheva.

The famed Congress Avenue Bridge in Austin, Texas, has a monumental population of Mexican free-tail bats. Every evening, as many as a million and a half of these insectivorous vacuum cleaners rise over the bridge like a smoke cloud (except during the coolest part of the year when they head south to Mexico). It is one of the largest populations of this species in existence, dwarfing the famed colony of Mexican free-tail bats that lives in Carlsbad Caverns National Park. It is also the single largest population of bats in any North American city. Tourists love to watch them billow up in search of their nightly meals—a thousand tourists on a typical summer's eve. For that reason and perhaps also because Texans seem to have a natural affinity for anything really outsized, Austin loves them, too.

It was not always so. Twenty years ago, only a few thousand bats lived in the shelter of Congress Avenue Bridge. Then the city ordered its reconstruction. Engineers added expansion joints about an inch wide and 16 inches deep. Without ever intending it, they had thus designed crevices of just the right dimensions for Mexican free-tail bat roosts. Bat real estate agents were delighted. The roosts sold like dream homes and bats moved in by the tens of thousands.

And if you don't care much for bats, how about storks? No one meant any of the churches, town halls, or other buildings in Europe to be platforms for storks' nests. But that is what they look like to the storks. The towns lucky enough to host storks treasure them as if, all along, they had indeed built their belfries, turrets, and chimneys for storks. The locals often reinforce the huge stork nests with chicken wire. They census the stork populations and rejoice in the successful rearing of each new generation.

Happy accidents will not suffice to prevent mass extinction. But neither are they extraordinarily rare. We need to search for them and take advantage of them, because sometimes they turn up in the most unexpected circumstances. Imagine a military tank chewing up the land over which it passes. Next to an exploding bomb or a cloud of Agent Orange,

Storks nesting on an old church in Ejea de los Caballeros, Spain. Photo by Sanda Kaufman.

that tank must be the most improbable source of ecological benefit you can think of. But if it happens to be on a military base in Indiantown Gap, Pennsylvania, think again. The regal fritillary actually benefits from that tank.

The regal fritillary is a butterfly. Never common, it lives in eastern North America in a narrow strip of territory stretching from New England to Nebraska. The strip lies astride the isotherm of 50 degrees Fahrenheit—that is, the regal needs places whose average annual temperatures are close to that. But who cares where regal fritillaries like to set their thermostats? What I want to know is why they welcome tank training exercises? The answer is pretty simple.

The caterpillars of regal fritillaries hibernate in the soil. Females lay their eggs in late August and September. After three or four weeks, the eggs hatch and the tiny caterpillars go straight into hibernation. The torn up soil of tank-training areas makes an ideal place for them to find shelter over the winter. Tank caterpillar treads cultivate the soil in just the right way for planting tiny regal caterpillars.

Naturally, I am not recommending that you call up your local militia and ask them to favor your backyard or neighborhood park with their next tank maneuvers. But tank maneuvers do happen, and if they produce a silver lining for some species, we ought to give them credit. Not only that, but what's good for a few species might be made good for many.

A Tropical Tragedy

Soon after the Second World War, some serious businesspeople invaded the middle elevation tropics of Mexico. They cleared the forests, opening large tracts of land for cattle ranching. But that is not the tragedy. The tragedy is that—in at least one place—these ranchers do not raise cattle anymore.

The place I am thinking of is in Chiapas. It is called the Ocosingo Valley and it sits about 3,000 feet above sea level. In addition to its pastureland, the valley has extensive patches of managed and unmanaged woodlands—coffee plantations, remnant strips of forest along its two rivers, pine and oak woods, and patches of a species of acacia.

The acacia, *Acacia pennatula*, looks like it belongs in giraffe country. It is a small tree 20 to 35 feet tall, with a wide canopy. But it has far too many spines for cattle to eat it. So the ranchers employ people to eradi- cate acacia from their pastures. They keep it out by destroying whatever seedlings volunteer in the pastures. Otherwise it would overgrow the land and ruin their livelihoods. And yet, they do not utterly extirpate it from the valley. Why?

They need it to feed their cattle. Although the stems and branches are too spiny to eat, the pods contain large amounts of protein and make great cattle food. After the pods drop from the trees in the dry season, the ranchers move their cattle into woodlots composed almost exclu- sively of acacia, where the cattle chow down.

The ranchers also need the acacia for fence posts. This is hardly a minor requirement in a valley where a fence post lasts only three or four years. No one wants their valuable animals to wander off.

And so the Ocosingo Valley retains a number of acacia woodlots.

Russell Greenberg and his colleagues decided to see if these woodlots support birds.[9] They looked for birds in 18 different Chiapas habitats, from pastureland to lowland rainforest, from coffee plantations to high elevation pine forests. And, naturally, they surveyed the acacia woodlots. No one expected what they found there.

The acacia woodlots turned out to be havens for a large number of songbirds that migrate to Mexico from the north. They have more spe- cies (18) and more individuals than any other habitat—including "natu- ral" forest at low elevation. Nine warbler species make it their winter resort. Hummingbirds, vireos, and finches use its abundance. (By the way, in second and third places are shade coffee plantations and rem- nant forests along rivers—both also habitats associated with human use of the land.) The way the ranchers of the Ocosingo manage their woodlots provides a splendid case of reconciliation by accident.

Unfortunately, Chiapas lies at the heart of a political dispute that some- times gets lethally violent. In the past few years, the Ocosingo Valley's ranchers have given up and moved their cattle out. Even the biologist- researchers have had to retreat. Sadly, the birds have no place else to go.

The deepest tragedies have morals. This one provides an instructive example. Reconciliation may happen accidentally, but no one can de- pend on the accident to keep happening. We need to recognize our happy accidents and resolve to keep doing the right thing.

CHAPTER 8

The Tyranny of Space

Earth so huge, and yet so bounded—
pools of salt, and plots of land—
Shallow skin of green and azure—
chains of mountain, grains of sand!

Alfred, Lord Tennyson[1]

Many different sorts of people care about conserving the world's species. Many of them influence conservation, too, and most are not biologists. The practice of conservation depends crucially on policy makers, politicians, economists, and engineers. Moreover, an army of concerned volunteers press on the body politic, giving of their time and their substance.

This makes the practice of conservation quite unusual. I know of no other branch of biological science that so involves laypeople in its front lines. You would not ask a pharmaceutical house to employ nonbiologists in its research and development. You would not go to patients to run the clinical tests that evaluate new drugs.

To be effective, our volunteers, our citizen-conservationists, must be committed. To be committed, they must believe. However, reconciliation ecology is neither a religion nor a political philosophy. If it were, I would preach it. But, at its core, it is a branch of science. So the citizen-conservationist must first understand it in order to believe it.

In this chapter and the next, I will outline the science of the problem that diversity faces. Then I will explain some of the scientific results and show you how they predict the future. Soon you will come to see that reconciliation ecology is more than an interesting alternative to other strategies of conservation. Reconciliation is a must.

Species-Area Relationships

While George Washington was serving his second term as president of the United States, a minor German noblewoman took to her deathbed in Prussia. When she died in November 1796, she left behind two things that would revolutionize natural history and ecology. One was an inheritance appropriate to her aristocratic station. The other was a son who would focus that inheritance on his dream. Because of his single-mindedness, that dream would spread, first among scientists, then among us all. His name was Alexander von Humboldt, or more formally, Baron Friederich Heinrich Alexander von Humboldt.

In 1797, supported by his inheritance, Baron von Humboldt resigned his position as mining engineer at Steben (near Bayreuth) and set out on several unusual journeys in Europe. He went not to visit some spa or be seen in the right palace, but to study scientific phenomena that are bounded by time and space. You see, a molecule of water is a molecule of water, no matter where and when you encounter it. You can study a representative molecule of water wherever you live. But if you want to examine, say, the courtship rituals of great bustards, you had better head for someplace where great bustards live, like the open fields of Spain. Not that Humboldt was so interested in ornithology. He wasn't. But he did hanker after rocks and plants and folkways. You don't study such things from your own apartment.

Obvious, you say? Well, most obvious things take genius to reveal and this was one. Humboldt's invention of "scientific traveling" influenced Charles Darwin and Alfred Russell Wallace and scores of other nineteenth-century luminaries. It proved to be the tool we needed to close the books on sea monsters and unicorns and get on with the scientific description and analysis of the world of life and the lives of peoples. Scientific traveling sparked a scientific revolution.

Humboldt's traveling yielded discoveries that mushroomed well beyond the many thick volumes of particular details that he spent his fortune to observe and publish. Despite his lust for detail, his first love was

102

synthesis and he was remarkably good at it. He invented weather maps; he discovered that the Earth's magnetic force decreases from the poles to the equator; and he perceived and proclaimed the most general patterns known to ecology.

For example, did you know that in 1800, scientists thought all the world's tropics had much the same few species of plants? Why? Because every tropical plantation lay close to an ocean over which it could send its produce and get resupplied. Using their marine trade routes, tropical plantations had homogenized themselves and their surroundings. They had imported to all tropical plantations a certain few plant species that people wanted or needed So all of them, on every continent, shared the same few species of plants. Meanwhile, Europeans who lived in the tropics stayed close to some kind of plantation, and no naturalist sent to accompany a voyage of exploration ever strayed far from his ship. That explains why those few plant species were almost the only tropical plants known.

Humboldt rebelled and penetrated the interior. Under extraordinarily trying conditions—he had no helicopters or land rovers, no insect repellents or bottled drinking water—he literally blazed a trail deep into South America. Wherever he went, he collected plant specimens and studied native institutions (he was also a pioneer ethnographer). And he discovered that tropical America overflows with species—an immense number of species—seen nowhere else.

Even after two centuries, Humboldt's own words resonate with the awe of discovery:

> From the banks of the Orinoco to those of the Amazon and the Ucayali, over an extent greater than 500 leagues (2500 km), the entire surface of the earth is covered with dense forests; and if the rivers did not interrupt their continuity, the monkeys could, by springing from branch to branch, travel from the northern hemisphere to the southern hemisphere. But these vast forests do not present a uniform appearance; each portion produces a variety of kinds . . . no plant dominates the others.[2]

Thus Humboldt taught us that the tropics are the Earth's most diverse regions. His discovery has burrowed so deeply into the Western mind that everyone now takes it for granted. I cannot imagine a world ignorant of the image of a majestic tropical rainforest teeming with diverse life.

A second Humboldt pattern dominates the science of species diversity: The number of species native to a region depends on the region's size.[3] In other words, the bigger the region, the more species it has. Today's ecologists call this pattern "the species-area relationship."

Few nonecologists know about the species-area relationship. Yet ecologists take it as much for granted as they do extravagant tropical diversity. In the end, you will see that the two patterns are intimately connected.

What Species-Area Relationships Look Like

Humboldt discovered a pattern but we really cannot make much use of it until we can describe it better. Professor Hewett Cottrell Watson, premier botanist of Great Britain during the early nineteenth century, was the first to try. With his encyclopedic knowledge of British plants, he systematically examined how many species lived in various areas. He noticed that, indeed, the diversity of plants did depend on the size of the area. But even more impressive, he noticed a regularity in the

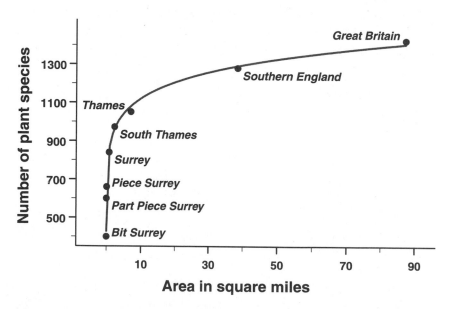

Number of plant species in Great Britain and successively smaller pieces of Great Britain. The smaller the area, the fewer plant species live in it. But the extreme curvature of the relationship makes it hard to see and to use, especially over the smaller areas (on the left side of the chart).

pattern: "On the average," he wrote in 1847, "a single county appears to contain nearly one half the total number of species in Britain; and it would, perhaps, not be a very erroneous guess to say that a single (square) mile contains half the species of a county."[4]

Now an average British county has about 1,250 square miles and the entire island of Britain has about 87,417 square miles. So Watson's pattern is a law of diminishing returns. The first square mile contains about 400 species, but adding a second square mile does not make the total 800. That requires a whole county, that is, about 1,250 square miles. In other words, it takes about 1,249 square miles to add the second 400. And the returns keep on diminishing because although one county has 800 species, two counties do not contain 1,600 species. To find 1,600, a botanist will have to search out almost the whole island. It takes nearly an added 86,167 square miles to double the job done by the first 1,250.

The law of diminishing returns causes the trend line in the chart of Professor Watson's plants to curve convexly up; over large areas, added area brings very little return of added species. But a curved line is hard to use for predicting anything, diversity included. For instance, suppose you wanted to know how many plant species you would expect to find in 500,000 square miles. First you would have to find a formula to fit Watson's curve. Only then could you use the formula to calculate your prediction. That's not so easy, even today. Curves take a lot of math.

To progress, we need trend lines that are straight. Perhaps this appears impossible, but it is not. Ecologists get straight trend lines simply by converting the numbers in the chart to logarithms (or logs). What are logs?

We can write the number 100 as 10^2—that is, ten squared. The number 2 is called the log of 100. Now imagine a series of numbers: 1, 10, 100, 1000, and 10000. The logs of 1, 10, 100, 1000, and 10000 are simply 0, 1, 2, 3, and 4. So we can write this same series as: 10^0, 10^1, 10^2, 10^3, 10^4.

We can write intermediate numbers with logs too. For instance, a number lying between 10^2 and 10^3—say 240—has a log between 2 and 3. In fact, $240=10^{2.38}$. Similarly, the log of 20—which is a number between 10^1 and 10^2—is 1.30; $20=10^{1.3}$. Let us add these to our series:

Numbers:	1	10	20	100	240	1000	10000
Logs:	*0*	*1*	*1.3*	*2*	*2.38*	*3*	*4*

So, logarithms are just a handy way to rescale numbers. The bigger the number, the bigger its log. There are no exceptions.

Plants of various parts of Great Britain

Place	Square miles		Plant species	
	Simple	Log	Simple	Log
Bit Surrey	1 →	0.00	400 →	2.60
Part Piece Surrey	10 →	1.00	600 →	2.78
Piece Surrey	60 →	1.78	660 →	2.82
Surrey	760 →	2.88	840 →	2.92
South Thames	2316 →	3.36	972 →	2.99
Thames	7007 →	3.85	1051 →	3.02
Southern England	38474 →	4.59	1280 →	3.11
Great Britain	87417 →	4.94	1425 →	3.15

Moreover, if we convert a number to its logarithm, we lose nothing. We can always convert back with a calculator. Thus if we can predict the logarithms of species diversities, we will have simultaneously predicted the diversities themselves.

Now we can use the rescaling magic of logarithms to straighten out Professor Watson's curve. The table lists the places in his chart, their areas, and the number of plant species they contain. Also, next to each

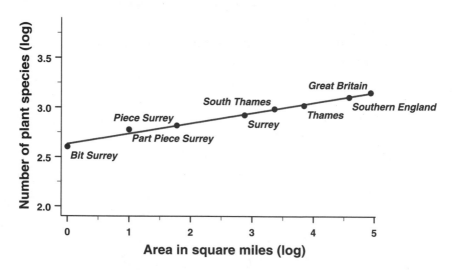

Exactly the same species-area curve of British plants as in the previous chart, but this time charted as logarithms. Logs transform the curve to a straight line. Its slope is 10.5 percent.

106

area and each plant diversity is its logarithm. We plot the logarithms in a second Watson chart and examine the result. No doubt about it, the trend line is now straight.

The slope of a line is the same thing as the grade you face when driving a mountain road. Straight lines have a constant slope. That is what makes them easier to work with than curves. In the case of Watson's plants, there is a 10.5 percent grade. Uphill.

This slope gives us something powerful to work with. In particular, it tells how much *log*-diversity bang we get for an added *log*-area buck. (The "bucks" are the area units of our chart. The "bangs" are the species diversity units of the chart's vertical axis. Moving, say, from a *log*-area of 3 to a *log*-area of 4 is increasing by one unit of *log*-area; so is moving from 2.2 to 3.2, etc.) The 10.5 percent slope in Watson's second chart tells us that for every added unit of *log*-area, we get an additional 0.105 units of *log*-diversity. Presto. There is our diversity prediction! The law of diminishing returns seen in the regular chart has changed to a law of constant returns in the logarithmic chart.

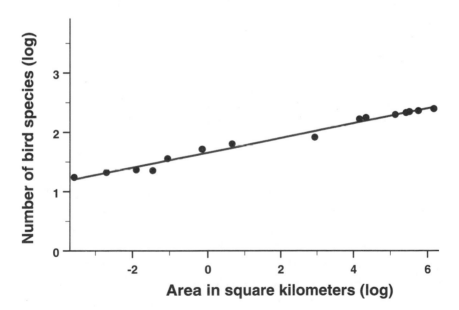

The species-area curve for California's chaparral birds. Its slope is 12.4 percent, very much like that of the British plants, and very much like those of a wide variety of species-area curves for all sorts of plants and animals in places all over the world. The trick to getting such low slopes is to stay inside a single continent.

Are Professor Watson's plants special? No, they are the rule. Imagine that we leave Britain to look at the plants in Ecuador or examine some other form of life besides plants. Regardless of what we look at and regardless of where we look at it, diversity will rise with area, and it will rise with diminishing returns. Furthermore, we will almost always find that logs clarify, quantify, and straighten out von Humboldt's species-area relationship. Birds and ants. Flowers and ferns. It does not seem to matter. Nor does it matter where we count them. North America. Amazonia. Australia. France. Israel. No problems anywhere. The chart of bird diversities in California chaparral shows another example.

Now comes the most amazing and useful fact: The *log*-diversity bang doesn't depend all that much on what you study or where you study it. Independent of environment and the kind of life studied, the slope hovers between 0.1 and 0.2.

But there is one hitch. Like Watson, you have to count your species within a single region, like Britain or California. Even a region as large as Europe is OK. But you must not compare different islands or different continents. If you do, the logs will still give you a straight line, but the species-area trend lines from different islands or different continents always have a slope higher than 0.2. That amounts to another rule. In the next section, we will make it fairly precise.

Not One, but Three Species-Area Relationships

Suppose we compare different continents by measuring their areas and counting their species. How much bang does *log*-area pack then? The slope among continents turns out to be at least 0.6 (60 percent) and generally close to 1.0 (100 percent). Check the charts to see two examples of steep and straight intercontinental relationships.

Now suppose we compare different islands of a single archipelago by measuring their areas, counting their species and calculating their species-area slopes. Most often the slope will equal about 0.3, but occasionally it will wander up as high as 0.55. That gives us another important rule: *Island slopes have intermediate values.* An archipelago always yields a bigger slope than those that come from within regions, and a smaller one than those that come from separate continents. The charts show a couple of examples.

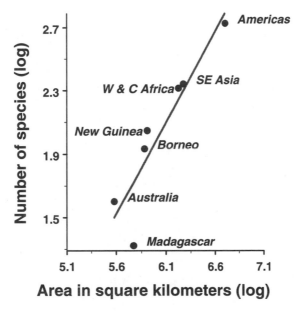

The species-area curve for fruit-eating birds and mammals living in the tropical forests of different continents. Its slope is 115 percent, roughly ten times as steep as those from within continents.

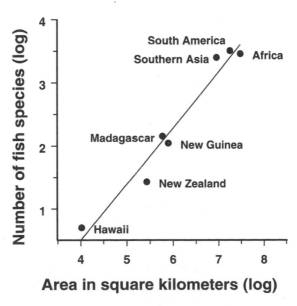

The species-area curve for fresh-water fishes of different, mainly tropical, biological provinces. Its slope is 89 percent. Thanks to Peter Reinthal for the data.

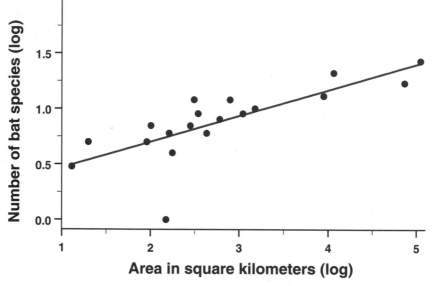

Bat species of different Caribbean Islands follow a species-area curve with a slope of 23 percent. Island slopes fall in between those from separate continents and those from patches within a single continent.

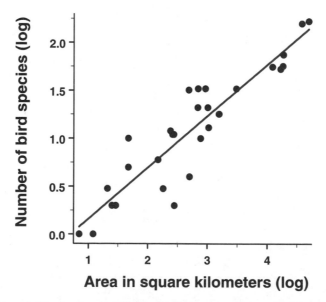

Bird species of different Tropical Pacific archipelagos follow a species-area curve with a slope of 54 percent. This is at the high end of island values, but still below those for separate continents. To produce these points, Greg Adler summed the areas of each island in an archipelago and listed all its species. (G. H. Adler, 1992. Endemism in birds of tropical Pacific islands. *Evolutionary Ecology* 6: 296–306.)

Let's list the rules that descend from von Humboldt's observation:

- Larger areas harbor more species.

- Adding more and more area leads to diminishing returns of new species.

- Keeping our books using logarithms presses the relationship into a straight line.

- The slope of this line is about the same regardless of what kinds of plants or animals we count or where we count them.

But the slope does depend on the way we select the patches we will count:

- If we study patches within a single region, the slope lies between 0.1 and 0.2.

- If our patches are different islands of an archipelago, the slope lies between 0.25 and 0.55.

- If our patches are different continents, the slope lies between 0.60 and 1.15.

These rules are ecological laws. They are nature's laws. They are God's laws. A few places get away with bending the law a little—some points in the charts lie above the line. And other places aren't treated quite fairly—their points lie somewhat below the line. But no place deviates very far. No place really flouts the law, and none falls too far beneath its proper station on the trend line.

The Causes of Species-Area Laws

Merely knowing the species-area laws cannot be allowed to satisfy us. If we are going to depend on them to predict how much land to set aside for wild species, we need to have confidence that they will last, that they really are laws and not just ephemeral statistical results. That seems vital to discover in the face of the profound changes that we are visiting on our planet. How can we determine whether those changes alter or even abolish species-area laws? If all we can do is recite them, we can never be sure that they are any more permanent than last year's list of favorite TV shows. I believe that we can gain confidence in species-area laws only by seeking their causes and asking whether those causes are likely to remain powerful in this new world of ours.

Here is the general framework that will organize our search for causes: Nature is dynamic in every area, no matter what its size. New species appear; old ones disappear. Diversity is the net result of Nature's dynamic give and take. Our search will therefore focus on the processes that add or subtract species.

Species-area relationships among continents

I begin at Nature's most majestic scale, entire continents. Here Nature's "give" is the actual birth of a new species. Her "take" is utter extinction. If we are going to understand why species diversity grows with continent size, we will have to explain how continental area controls the rates of speciation and extinction.

Extinction: Population is continental area's first link to extinction. Area supports population by supplying food and shelter. So, the more area a species inhabits, the larger its total population.* Since extinction might come about by means of the accidental death of every single individual in a species, large populations help to protect a species from extinction. The idea is simple: Think of a species as a life insurance company. A life insurance company that has only three clients, Tom, Dick, and Harry, will go bankrupt sooner than one that insures *every* Tom, Dick, and Harry.

---

*This is a strong tendency, but not as certain as death. For instance, I know very well that not all species whose individuals are spread out widely over most of the North American continent have larger populations than all species restricted, say, to a single mountain range in northern Mexico. For a specific example, consider a big raptor like the golden eagle. It ranges over the entire northern hemisphere—Asia, Europe and America! Yet, I doubt there are as many golden eagles as there are *Glossina morsitans*, perhaps the best known of the tsetse flies. Despite their widespread evil reputation— tsetse flies transmit African sleeping sickness—these tsetse flies live in only a few countries of tropical East Africa. I've thought a lot about such ifs, ands, and buts. However, I hope to present you with a clear picture of what I believe—very strongly—is a clear situation. So you'll hear no more about tsetse flies from me.

A second link between area and extinction stems from what ecologists call the "metapopulation view." According to this view, we ought to look at a species as a set of semi-isolated populations, rather like palms inhabiting the oasis patches of a desert. Each unit of the metapopulation could disappear by accident. Maybe a cold front slips too far toward the equator, or a hurricane devastates an extent of seacoast. The result would be an empty patch. But eventually some individuals—largely by accident no doubt—would find and recolonize the patch and re-establish the species's presence in it. Extinction would occur, however, if by chance all the metapopulation's units disappeared at once.

Now we can understand how the metapopulation view connects area to extinction rate. A species spread over a large area will tend to have more units in its metapopulation. It is not very likely that all will disappear simultaneously. That searing drought which covers enough area to eliminate five units of the species will not get them all. But a species restricted to a small area may not be so lucky.

Whether viewed in the population or in the metapopulation framework, area defends species against extinction. If a big continent and a little one both have—say—1,000 species, those on the little one will wink out faster than those on the big one.

Speciation: Most species are produced by a process called "geographical speciation." Geographical speciation begins when some geographical barrier subdivides a single population into a number of isolated subunits. A canyon can do that, or an ocean, or a mountain range. Gradually, the genes of each of the isolates change. But because of the barrier, the isolates cannot exchange genes very often; so their genes become more and more dissimilar. Eventually, if the barrier lasts long enough, each isolate founds a new species.

Now recall that a species living in a larger area probably has more units in its metapopulation. Under the right circumstances, these can be the isolates that will begin new species. So, a larger area is likely to have more species buds, metapopulation units whose separation is so complete as to allow the foundation of new species. That explains why a larger continent with 1,000 species should have a faster speciation rate than a smaller continent with 1,000 species. Though area slows the extinction of species, it speeds the dominant process that produces them.

Establishing a balance: What keeps diversity within limits? Why doesn't it grow boundlessly or disappear utterly? The general answer is that

113

speciation rates and extinction rates tend to balance each other because both extinction and speciation are feedback processes.

In a feedback process, the amount of what is being produced affects the rate at which it is being produced. What makes speciation a feedback process? Species are the factories that make new species. What makes extinction a feedback process? The more species there are, the more species there are at risk of extinction.

Feedback processes often allow a system to regulate itself. Those that affect diversity are good examples. As diversity grows, each species tends to lose part of its geographical area to new species. As its geographical area declines, so does its rate of speciation. But the same loss of area raises its rate of extinction. At some level of diversity, the rates balance. Nature's give equals Nature's take.

To understand the dynamics better, let's make an analogy to a leaky bucket. Think of a continent as the bucket and diversity as water. Speciation pours in the water while extinction, represented by holes up and down the side of the bucket, lets it leak out. To simplify the analogy, imagine that speciation pours in the water at some constant rate. In the beginning, most of the holes don't matter. They are above the water line. So, the water rises. As it does, however, water begins pouring out of more of the holes, increasing the leakage rate. Soon the combined leaks lower the water line as fast as the bucket is being filled. The water level stabilizes. Mimicking such a process rather closely, species diversity also seeks an equilibrium value on each continent.*

How many species constitute the equilibrium? It depends on the size and number of the leaks. The more leaks and the bigger they are, the less the water will rise before it reaches its sustainable level. It also depends on the rate at which we pour in the water. The faster the inflow, the higher the sustainable water level. Big continents have smaller, fewer leaks and take in water faster than small continents. That explains why big continents tend to have more species than smaller ones do.[5] It's also a major part of why the tropics abound in diversity. Tropical ecosystems are the world's largest land-based ecosystems.

---

*I could make the analogy more complex and precise but there is no need. Water in the more complex bucket would still reach a sustainable level.

The number of species in a continent is like the amount of water in a leaky bucket. The water leaking out represents extinction. The water flowing in symbolizes the generation of new species.

Here is why the amount of water in the bucket reaches a balance (called a steady-state): If the water level is very high, then water will be leaking out of most of the many holes and the inflow will not be enough to maintain the high water level. But if there is little water, it will be leaking out of few holes, and the inflow will be enough to raise the water level. Result: a balance between the rates and a steady-state amount of water in the bucket.

The steady-state diversity of a continent will grow larger if speciation rates increase (water flows in faster), or if the bucket has fewer, smaller holes (water leaks out more slowly). Perpetual water supply based on *Waterfall*, a lithograph of M. C. Escher (1961). Drawn by Greg Cooke, Arizona House of Graphics. © 2002, Evolutionary Ecology Ltd.

Habitat diversity

The previous section laid bare the causes of intercontinental species-area relationships but ignored both those within a single continent and those of islands. To deal with these, we must first consider how habitats come to be different from each other.

Once the interplay of speciation and extinction has given a continent its quota of species, those species can and do evolve. Natural selection forces them to adapt not just to their surroundings but to each other. Biologists call this process "coevolution."

In particular, when species find themselves together on a continent competing for limited space or food or nitrogen or the like, natural selection emphasizes their strengths. If they do well in a particular kind of soil, they will do best to restrict themselves to it as much as possible. If they get the most nutritive kick from a particular species of prey, they often do best if they pay less attention to other prey. If poker's your game, play poker and turn down the invitations to play gin.

Imagine the opposite. Say that a lion decided to hunt mice instead of antelope. Although a lion can certainly catch a mouse, it is not particularly well equipped to do so. The lion is too large to follow the mouse into its hole. Moreover, if you are a lion, you have this big, heavy body to feed. One mouse won't help much. So lions must hunt bigger prey. Furthermore, their big bodies help them to subdue bigger prey, adding value to their size, and perhaps putting pressure on them to evolve to be larger and even less suited for mousing.

Much the same thing happens when different species face a hodgepodge of environmental conditions. One species will—merely by accident—be better off here. Another will be better off there. Natural selection will see to it that each species limits itself to places where it does best and avoids challenging other species where those other species reign. Thus, the more species, the more narrow the range of conditions in which each species will be distributed. There's no magic in that. It is just the world of chance working together with the process of natural selection.

Now natural selection reinforces the environmental separation between species. It teaches each one how to specialize, that is, how to be even better at doing what it does best. It also teaches each species to spend even more of its time in the places where it excels.

We perceive the end product as different habitats. Red foxes spend most of their time in fields, grey foxes in forest, arctic foxes in tundra. For that matter, the weeds and grasses and shrubs that grow in what we call fields are also making a choice based on underlying values of climate and soil. Natural selection adjusts their seeds so that they tend to germinate in places (and at times) that give them the greatest opportunity to survive and reproduce. In short, habitat differences are there because life evolves to recognize them.

For most of my career, I had the opposite idea: Habitat differences are out there to begin with. I believed that God made them for the world of life using the tools of physics and chemistry. I believed that habitats form the mold that shapes species and what they do in ecosystems. Species meet the habitats, evolve, and specialize on them.

Forget that. We cannot meaningfully speak of different habitats except as mirrors of the specializations of species. God does not use physics and chemistry to differentiate habitats. God uses biology.

I had my eyes opened in several ways. My favorite happened on a mid-winter's day Down Under, 19 July 1993. Carole and I had stolen away for two weeks to visit Western Australia. What a remarkable corner of the world!

A habitat called the *kwongan* had given me my primary excuse for the visit. The kwongan is legendary among plant ecologists because it has an enormous number of plant species, species that occur nowhere else, not even in the rest of Australia. About 8,500 species of flowering plants live in an area roughly the size of South Carolina. More than 80 percent of them live only in the kwongan.

What do you imagine the kwongan looks like? You probably know that Australia has a lot of tropical forest, and that about half of Western Australia is tropical. So my guess is that you are thinking lush . . . dreaming verdant . . . picturing a rainforest perhaps. Not at all.

The kwongan is not even tropical. It sits in the southern half of the state and extends to the cool coast where nothing but the Southern Ocean separates it from Antarctica. The kwongan is not a tropical forest but an unlovely shrubland. To the untrained eye, it looks much the same as any scrubby, heathy Mediterranean shrubland in Corsica or Israel or California or Chile. But it isn't. It has two or three times as many species as they do. Only the fynbos, the shrubland of the Western Cape Province

117

in South Africa, can match it. The kwongan is not particularly large either, so it is one of those places that bend the laws of the species-area relationship. Doing my own little bit of scientific traveling, I had come to Western Australia with my wife to see this marvel and find out what makes it tick.

After several years of trying and after another bout of scientific traveling (this one to South Africa), my colleagues and I cracked the puzzle.[6] The kwongan and fynbos are so rich in plant species because their very poor soil gives an ecological advantage to shrubs with very short life spans. And the very short life spans speed up speciation rates. Hence, in the kwongan and the fynbos, nature's take is normal but her give is unusually generous.

But what truly concerns us here is the remarkable landscape. Shrubs on sand plain. Almost unbelievable! Flat, ridiculously impoverished sand covered with a riot of shrubby species, each one looking much the same as any other. For instance, the kwongan has several hundred species of *Acacia*. None of these are the tall trees you see giraffes nibbling on when

Shrubs on sand plain: A view in the *kwongan*, Kalbarri National Park, Western Australia. © 2002, Evolutionary Ecology Ltd.

118

you watch a film on African savannahs. Instead, they are all little more than overgrown shrubs. There are no trees in the kwongan, and precious few herbaceous flowers.

Let me give you a notion of just how monotonous this landscape is. In our early fifties, three years away from grandparenthood and totally unprepared for a challenging hike, Carole and I nevertheless decided to climb down the storied Murchison River Gorge, the deepest canyon in the Kalbarri National Park, heart of the kwongan's subtropical northern end. Though it was a weekday, several Australians had also journeyed to the canyon to make the descent. Calmed by their presence, we began our trek.

It took us about 15 minutes to get all the way down.

Oh, and it took about 20 to climb back out.

As I said before, this is flat country.

And the soil is pitifully poor all over. Leached poor by the rains of 350 million years. It takes a fancy piece of scientific apparatus to detect the tiny concentrations of nutrients like potassium and nitrogen that remain. Yet the plants have adapted. In fact, they actually require the poor soil. One of the surest ways to kill them is to fertilize it!

Clearly, no human would look at the monotony of the landscape and say that it has a lot of habitats to fill. Yet those plants appear to have done the unthinkable. They have looked at that monotony through the eyes of natural selection and seen a great variety of soils. They have specialized on the minuscule differences of soil nutrients from place to place in the kwongan. Yes, although at first we cannot see it, the kwongan is a kaleidoscope of habitats. That's what the plants say, and theirs is the only opinion that counts.

So, speciation and extinction give a continent (or some large region of it like the kwongan) a sustainable diversity. Then the species evolve specializations that restrict the places they live in, thus subdividing their world into what we see as different habitats.

Species-area relationships within a region

Now we can explain Professor Watson's species-area pattern. If you study a small area inside a region, you will see only a few of its habitats and only species that can live in those habitats. As you increase the scope of your observations, you can expect to see more and more of the region's habitats plus their special species. More area, more habitats; more habitats, more species. That is the cause of Watson's species-area pattern.[7]

Why do logs fit the pattern so well? Answering that question takes a long, difficult mathematical argument, but it has been done.[8] It shows that we should not ascribe very much biological significance to the fact that logs tend to press species-area relationships into straight lines. The reason lies not in biology but in the mathematical theory of probabilities.

And why should that straight line have a slope between 0.1 and 0.2 within a region? That question gave ecology fits for over 40 years, but very recently, Brian McGill, who is my doctoral student, solved it.[9] The answer flows from a few basic facts about different species:

- Species contain different numbers of individuals.

- To persist, species need a minimum number of individuals.

- Some species live in only a few restricted places, others are more widespread, and still others live almost everywhere.

- Where one species lives (its "geographical range") does not depend on where any other lives.

- Each species exists below its average population density in most of its range.

None of these facts is in dispute and none seems likely to change. So McGill's mathematical explanation gives us some assurance that Professor Watson's line is going to stick around for a long time.

Species-area relationships within an archipelago

Consider the species diversity of Ceram, Indonesia. It lies off the northwest coast of Oceania (Australia-New Guinea) about 100 miles from West Irian, Indonesia. (West Irian is a large piece of the island of New Guinea.) Ceram is a true biological island because on it, Nature's give is the rate at which species find their way to Ceram from someplace else (mostly from New Guinea in Ceram's case). Nature's give will not be Ceram's speciation rate—that's only a trickle. All evidence indicates that the flood of species' immigration to a real island like Ceram overwhelms any trickle of speciation it may experience.

Nonetheless, islands too have a sustainable diversity. Here is the logic: If all the species in its region have already immigrated to an island and live there, then Nature has nothing left to give. The species' immigration rate will be zero. So we have to change our water-bucket analogy a bit. We cannot imagine water pouring in at a constant rate anymore. It must pour in fastest if no species has yet found the island. As more and more do, the immigration rate must fall to reach zero after all species have arrived. Meanwhile the extinction holes in the bucket's sides are leaking species just as before. As more species come to live on the island, the water rises and water flows from holes higher and higher in the bucket's sides. Eventually, the increasing extinction rate balances the declining immigration rate. Result: island equilibrium.[10]

Now we extend our thinking to a set of islands close to Ceram, but of varying size. This is readily done in the Sunda archipelago of Indonesia and her neighbors. Because they are all a similar distance from New Guinea, they should have about the same rates of immigration (provided we start them at the same species diversity). But they will not share extinction rates. Compared to small islands, bigger islands will have more habitats, and these will protect more kinds of newcomers from extinction. Bigger islands will also have larger populations, and these will keep extinction rates lower too. With smaller, fewer holes in its bucket, a bigger island will have a greater sustainable diversity than a smaller one. Result: a species-area relationship among the islands.

What explains the differences in species-area slopes?

Brian McGill's math accounts for the small slopes of species-area relationships within a region. But it does not tell us about the slopes of species-area relationships at other scales. Why should the slopes among islands range from 0.2 to 0.6? Why should the slopes among continents range from 0.6 to 1.0? Sorry, folks. The jury's still out. We are not yet sure how to account for the actual values of the slopes. But we can show

121

that intercontinental slopes must be the steepest and that island slopes must be intermediate.

Let's imagine four pieces of territory. One will be a large continent, say Oceania. Another will be a tiny continent. Hawaii will do beautifully, and I will explain why in the next several paragraphs. Now, we need an island about the same size as our tiny continent. The five principal islands of Hawaii constitute its chief area and add up to 6,039 square miles. Ceram is a good match at 6,046 square miles and, like Hawaii, is also tropical. Finally, we need a 6,000 square-mile tropical piece of the continent of Oceania. We may as well take it from West Irian.

Now I have to explain my apparently inconsistent use of the terms island and continent. Everybody knows that Hawaii is not a continent but an archipelago. And New Guinea is an island, too. So, how can West Irian be part of a continent?

Usually, we let the geographer tell us what is an island and what is a continent. But sometimes even the geographer has a problem. Is Australia a huge island or a continent? And sometimes the biologist sees interesting things the geographer may not care much about, like population processes or evolutionary histories. Then the biologist needs to assert himself. He may need to insist that what looks like an island to a geographer is actually a continent to the world of life.

What justifies calling Hawaii and New Guinea continental? Just this: They both produce almost all their own species. They are both fountains of diversity. So when we consider how much water they have in their buckets, we will have to ask questions about their speciation and extinction rates, just like we do when we consider Africa or South America. In contrast, a true biological island, like Ceram, adds species by immigration.

Now we return to our puzzle. We wanted to figure out why intercontinental slopes are steepest, slopes from inside a region are shallowest, and inter-island slopes lie in between. We chose four territories—the large continent of Oceania, the mini-continent of Hawaii, the island of Ceram and a piece of West Irian. Oceania has 3.3 million square miles (the log of 3.3 million is 6.52). Each of the other three has about 6,000 square miles (the log of 6,000 is 3.78). But you will now see why the three areas of 6,000 square miles should contain quite different numbers of species.

For our example, we will consider the species in two subfamilies of ants.* E. O. Wilson reported 10 species of these ants on Ceram, and 126 on the entire island of New Guinea.[11] Based on his samples of pieces of New Guinea, we can estimate that 6,000 square miles of West Irian have about 81 species† and that Oceania's 3.3 million square miles have about 142. In contrast, none of these ant species is native to Hawaii. None.

Let's consider those numbers carefully. What explains the diversity of the piece of West Irian? First, it has fewer species than all of Oceania because it is an incomplete sample of Oceania's ant habitats. Nevertheless, almost certainly, it has even more ants than those habitats can sustain by themselves.

Ants, you see, move around rather freely within a continent. For a short time in their lives, male and female ants have wings and fly about looking for a spot to establish a new colony. Caught up in the wind, they often export themselves inadvertently from places where they do well to places where they cannot quite make ends meet.

A continental place can regularly "import" such individuals. Their flow replenishes the stocks of its failing species. Thus, a piece of a continental area will seem to have healthy populations of species whose habitats it really does not contain. Population ecologists call perennially failing units of a metapopulation "sinks." The units in the metapopulation doing the exporting from outside the area are "sources." It's hard to tell the difference between the source and sink species without a great deal of prolonged census work. In practice, we rarely try. Instead, we put each species in our list of local ants without distinguishing between source and sink species.

---

*Perhaps you don't much care about the world's treasure of ant species. You may even be thinking that once you've seen one ant, you've seen them all. As Robert Cauthorn of the *Arizona Daily Star* wrote, "How many types of ants are really necessary to spoil your picnic?" But think of ants as guinea pigs. What we learn from them, each of us may be able to use on behalf of our own favorite life form. Besides, ants are the creatures of choice for E. O. Wilson. They've taught him quite a lot and inspired him to do good works on behalf of all life forms. So ants are a guinea pig and an inspiration, a light unto the species.

†These numbers tell only a fraction of the truth. Tropical ants weren't very well known in 1960 and ant-loving Australian biologists have speculated that their continent has over 10,000 species! Yet Wilson's old numbers do allow us to make good comparisons.

So our piece of West Irian will have some sink species and some source species. Some of its 81 species will really not have adequate habitat inside the area. If we were to surround it with an ant-proof barrier to prevent imports, the sink species would soon vanish from our list and only the source species would remain.

One hundred miles of open ocean is not a bad ant-proof barrier. So a well-isolated island like Ceram will have no sink species. It has only those species whose habitats it really does contain. Yes, we can expect Ceram to get an immigrant species now and then. But certainly not often enough to maintain a species whose average death rate exceeds its average birth rate. That is why Ceram has fewer species than a 6,000 square-mile area of West Irian. In fact, that is why all islands have fewer species than similar-sized areas of nearby continents. Islands lack sink species.

Now let's imagine that we can pull the immigration rug out from under Ceram. We might do this by floating it way out into the Pacific Ocean, way out, say, to the position of Hawaii. Virtually all additions to the species of such an isolated place must come from within. Internal additions are speciations. But the speciation process is slow, even slower than immigration. What will that do to Hawaii's list of species?

To figure it out, we revisit our leaky bucket. Moving it from Ceram to Hawaii has done nothing to fix any of its leaks. All we've done is replace the flow of immigrants with the trickle of speciation. The water cannot rise very high before the leak rate matches the trickle. So of course Hawaii has fewer ants. But it has none! Should it have zero?

We can check by putting the numbers in our *log*-chart format. The log of 6,000 is 3.78; the log of 3.3 million (Oceania's area) is 6.52. In the next chart, the three areas of 6,000 square miles line up vertically because they differ only in their species diversities.

We now draw two lines. The first connects the point representing all of Oceania to the one representing the piece of West Irian. Because this piece will harbor the most species of any of the 6,000 square mile areas, the line connecting it to all of Oceania must have the gentlest slope. The second line connects Ceram to Oceania. Ceram has fewer species, so its line is steeper. Island slopes are always going to be steeper.

With Hawaii, we run into a little construction problem. Hawaii has zero species and the log of zero is negative infinity. Negative infinity is infinitely below the bottom of the page on which the chart is printed, so we haven't got enough room to draw it.

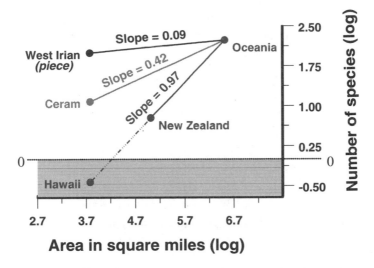

Hawaii has no native species of certain ants (specifically ponerine and cerapachyine ants). It is so far from a source of ant colonists that its ant diversity must rely mostly on speciation within the archipelago. For ants (and many other life forms) that means Hawaii is more like a continent than an island. Indeed, judging from the ant species-area curve for different continents, Hawaii is just too small to contain even one species; it falls below the zero value in the chart. One might say that there is not enough ant-water trickling into its bucket of native species.

But we're not entirely out of luck. We can estimate this line using bigger continents than Hawaii. And then we can ask that line a question: In view of Hawaii's size, is it reasonable for Hawaii to have zero species?

In addition to Oceania, we have comparable ant numbers for two other nearby biological continents: tropical, southeastern Asia and New Zealand (another geographical island whose species come from within). Suppose we shoot a line through these three points. For what area does it forecast a *log*-diversity of zero? That area will be the minimum area necessary for these ants because a log of zero translates to one species. Less than one is biologically equivalent to no species. Does the area of Hawaii fall below the minimum?

We extend the line joining Oceania and New Zealand southwest toward the *log*-area of Hawaii.* It has a slope of 0.97 and reaches Hawaii's area at a *log*-diversity of about –0.5. Thus the intercontinental species-

---

*Since our chart is getting a bit complex, I included only Oceania and New Zealand in it. Adding tropical southeast Asia makes only an imperceptible difference anyhow.

area relationship forecasts one-third of a species for Hawaii. No wonder it has none! According to our intercontinental line, the minimum area required to admit a continent to the ant club is about 20,000 square miles—6,000 just won't do. Of course, if, like Ceram, you are an island and you have a big continental buddy to back you up with a flow of immigrants, your price of admission gets deeply discounted.

The chart is complete. Separate continents have the steepest slopes and pieces of continents have the gentlest. Furthermore, though ants move around more easily than most species, every species can move to some extent. So our arguments apply to any kind of plant or animal that has trouble hopping about between islands or invading a new continent. That might not cover germs and molds, but it pretty well covers everything that matters to me.

Summary

Our examination of species-area patterns leaves us in good scientific shape. We now know that species-area laws exist. We also know that a set of dynamic processes (such as extinction and speciation) produce diversities that are sustainable and reflect the balance of Nature's give and take. Finally, we understand that we must look to different processes in order to explain the diversity of any particular area. The processes that dominate will be those with the fastest rates that apply to that area. In a whole continent, the fastest rates are those of speciation. On an island, the rate of speciation is overwhelmed by the rate at which species immigrate. And in a patch of a continent, immigration of species that are not present is overwhelmed by the rates at which individuals replenish the stocks of sink species. In the following two chapters, we will use these principles and the regularities of species-area relationships to predict the future of diversity on the planet.

CHAPTER 9

Falling Down the Time Shaft: The Case of the Incredible Shrinking Planet

If we see a light at the end of the tunnel,
It's the light of an oncoming train.

Robert Lowell [1]

The Earth today is a shrinking island adrift in the sea of time. Maybe you find that sentence somewhat mystical for a science book. But it is not really mystical. It is thoroughly scientific. This chapter has the job of explaining it. In what scientific sense is the Earth shrinking? And what makes it an island adrift in time?

By saying "shrinking," I do not mean some added consequence of global warming! It's not as if the Earth were made of wool or cotton, and we returned to the galactic laundromat to find a pea-sized planet in its dryer. But, for most species, that might as well be the case. Most forms of life survive only in natural environments. And we are shrinking those environments pitilessly. We convert them—square mile after square mile—to our own purposes. Moreover, whatever we take for ourselves, we change so radically that almost none of the other species in the world can still live in it.

So far, we have taken over about 95 percent of the Earth's land surface and refashioned it according to our own specifications. Jeremy Jackson maintains that the same is largely true in the sea: "Oceans are not wilderness and no Western Atlantic coastal habitat is pristine. The same is almost certainly true of coastal oceans worldwide."[2] For most species, then, it must seem indeed as if we left the hot water setting of the washer much too high. Their planet no longer fits them. They have naught to cover their nakedness, to shield them from the dispassionate harshness of nature.

The U.S. Departments of Interior and of Agriculture[3] paint a slightly rosier picture of the United States itself. They claim 10 percent of the land in the USA has "specially protected" status. They add that 90 percent of this is in the best set of categories: National Parks, wilderness areas, wildlife refuges, wild and scenic rivers. So, they claim, we've taken only 91 percent of natural U.S. lands instead of 95 percent.

But Noss and Cooperrider[4] point out that the 10 percent figure is quite misleading. In particular, we do not protect our wildlife refuges very well. Many are heavily grazed, logged, or hunted. And as for the national parks, well, in their eyes (mine, too) "Yosemite Valley . . . is as much a biodiversity reserve as Coney Island is a marine sanctuary."[5] Noss and Cooperrider admit no one knows exactly, but suggest that less than 3 percent is a truer estimate of what we have left.[6]

The situation varies in seriousness from environment to environment. For instance, we have yet to make devastating inroads in the world's coldest and driest places, her tundras and deserts. But, while such places are truly majestic and beautiful, they do not harbor most of the planet's diversity. By the same token, their very poor ability to support life makes them poor choices for our further expansion, and that protects them for now. I am not saying that the world's deserts and tundras have no ecological problems, but those problems generally have little to do with loss of area.

For exactly the same reasons that natural deserts remain vast stretches of territory with small human impact, the world's natural grasslands have nearly vanished. Grasslands are perfect places for us to grow our grains and graze our cattle. They are often flat and easy to plow. They tend to have fertile, deep soils. Their climates are moderately wet or moderately dry, moderately cool or moderately warm. In short, they make superb farmland and ranchland. Thus, they constitute the single environment most rare in anything like its natural state. *Less than one half of one percent* of the world's grasslands look like they did before humanity invented agriculture. If you were a tuft of prairie grass or an American bison, wouldn't you feel your world had shrunk?

Forests have fared a little better. For instance, the United States is particularly rich in forested area. In 1630, when European colonization of this country had just begun, forests covered about 423 million hectares. Today, forests still cover about 298.4 million hectares, which is a little more than 70 percent of their original area.

But a closer look at today's forests reveals that most are far from any natural state. Instead, they are forest plantations. Evidence from many parts of the world shows that plantations do not support much diversity. For instance, a forest plantation of Douglas fir in Oregon will support only 15 species of canopy insects and spiders, while a nearby old-growth forest of Douglas fir and western hemlock harbors 75.[7]

The trees of a forest plantation are more or less like vegetables, planted and harvested according to a carefully devised schedule. We specify even the genes of those trees. Forest plantations are just cornfields whose stalks have gotten very tall and turned to wood. They display none of the majesty of natural forests. It is hard for a vegetable to seem majestic.

So, what is really left? Not much. U.S. forest reserves contain our natural forests, and U.S. forest reserves of all types amount to only 19.2 million hectares.[8] That is merely 4.5 percent of what the Pilgrims started out with. Sounds moderate doesn't it? We take more than 95 percent and leave less than 5 percent. At least for now we leave it.

Moreover, some kinds of forest are very poorly represented in our natural forest reserves. For instance, eastern pine forests cover large tracts but most are plantations; pineland reserves cover only 495,000 hectares. That means only about 1.6 percent of eastern pinelands are natural forests. Forests of oak, gum, and cypress have similarly poor protection. Reserves make up only 201,000 hectares of 11.8 million; that's 1.7 percent in reserves.

And then we come to the tragedy of my generation. Where have all the rainforests gone? When I was a child, builders and furniture makers freely used luxurious mahogany woods from the Philippines, rosewoods from South America, teak from Indonesia, and ebonies from Africa. Dense, magnificently grained timber from the world's tropical rainforests. We took it for granted! Most people still do even if they can no longer afford anything more luxurious than a chipboard china closet. We cut down our planet's rainforests for their timber and cleared them to plant crops.

Chances are that you are troubled by the rapid continual disappearance of the world's tropical rainforests, now proceeding at the average rate of 160,000 square kilometers per year. Chances are that you have read some panicky newspaper report that such and such a stand was about to reach the end of its multimillion-year journey. Chances are that you have seen video footage of a forest being "harvested" or worse, cleared by a subsistence farmer with a lump in his throat, or by an agribusiness behemoth for a ranch or a farm with the same expected productive life span as a shopping center. If that.

But chances are that you have not heard about the worst case of all, the Atlantic Coastal Rainforest of Brazil. Sadly, there is very little left to hear about. Once, this forest covered about 12 percent of Brazil's area. Today, more than 93 percent has been entirely cleared, and most of the remainder is forest recovering from previous timbering episodes.[9] Yet, even that pitiful remnant retains enough majesty to set us weeping for what we have lost.

You see, Brazil is blessed with two massive sections of the great American tropical rainforest. You have certainly heard of its Amazon section, the very same forest region that so enchanted von Humboldt two centuries ago. But one particular hectare of the Amazon's sister, the less famous Atlantic Coastal Rainforest, has the highest number of tree species of any place on the face of the Earth. It is the Guinness-Book-record-holder, the mother of all hectares, the woodland of wonders. It has about 450 species of trees!

Four hundred fifty species in a single hectare.[10] What a phenomenon. Its nearest competitors—a hectare of Amazon forest and another one in Sarawak—have fewer than 250 species each. So this special hectare far outranks even its nearest rivals.

Let's put it in perspective. We humans live at most some 120 years. Imagine if a real person lived 200 years! Imagine if someone ran a mile in two minutes, or 100 meters in five seconds! Unthinkable? Now you appreciate what we have here. The entire United States has only 865 native tree species, and this slightly oversized house lot has 450!

Our champion hectare lies in the extreme southeastern area of Bahia State, a part of Brazil that God meant to be almost entirely rainforest. Over the years, biologists kept careful track of what has happened to the 2,724,700 hectares in this bit of Bahia. In 1945, it was still 85.4 percent forested, forest covering 2,235,900 hectares. Previous generations had removed forest from only a few places—in the neighborhood of a few towns, near the railway, along the seacoast and along eight rivers. Quite possibly, one of the hectares remaining in 1945 was even richer than today's supreme patch. But we will never know. From 1945 to 1960, 1,315,088 hectares of the remaining forest were cleared. From 1960 to 1974, an additional 605,812 hectares got chopped down. From 1974 to 1990, another 240,175 hectares were harvested. By 1990, fully 94 percent had been cleared of its tropical rainforest. Only 164,825 hectares remained. Maps[11] show that some of this logging actually took place inside the boundaries of reserves and national parkland set aside by the Brazilian people to protect what little they had left.

No doubt, the pressure continues to this day. Economic forces cannot be satisfied until all profitable hectares of these forests have been liquidated. Business people are driven by the demands of their jobs to turn in the best bottom lines. Most, I am sure, do not intend terricide. In truth, they cannot help themselves. They are slaves to unbridled capitalism, and society offers them little guidance or restraint. Society believes that jobs are at stake.

Yet if we stopped now, what would be so different? Eventually the primeval forest will be gone anyhow and the timber companies will shift to sustainable harvesting policies. But only companies that play by today's rules will live to make that shift. And today, those rules dictate that a company cannot compete if it restrains itself.

Perhaps we need to encourage companies to contract with each other to limit the devastation that unbridled competition forces them to perpetrate. Monopolistic? Yes, I'm afraid so. But maybe our natural heritage needs the protection of a limited few monopolistic practices. Anti-monopoly laws were enacted in the early twentieth century to protect the consumer by fostering competition. Who foresaw the possibility of a world where only certain monopolistic practices could save the world's last remaining primeval forests?

But that's another story. We cannot afford to dream too much. Instead, let us return to consider the effects of our actual policies, past and current. We need to ask what would happen if we stopped now and accepted virtually all the shrinkage that we and these policies have already caused. How can we make the best of such a bad lot?

How bad is it? At present, roughly 5 percent of the world's terrestrial habitat remains in something approximating its natural state. Not 10 percent. Not 1 percent. And not 5 percent of every kind of ecosystem either. As we have seen, deserts fare better, prairies much worse.

Very roughly 5 percent. That's about all the accuracy you can expect from a struggling ecologist these days! But, luckily, such accuracy turns out to be quite enough. We shall use it to give a rather robust prediction about the future of biodiversity. If our prediction errs by 10 percent or even 20 percent, it won't make much difference. Those who care, will still care. Those who don't, still won't. And our remedies will stay the same, too.

Does the shrinkage matter? Does it matter if only 5 percent of the world's natural areas remain? Why not just swear to take good care of the 5 percent? That's a lot of planet for wild things! We will live in the 95 percent; they will live in the 5 percent. It's a hell of a lot more space than Noah gave them in the ark.

Well, maybe so. But Noah did not expect his charges to remain on the ark very long. As we shall see, canny old Noah was taking advantage of a little known fact: Extinction is a process and it takes time—sometimes a very long time—from start to finish. Let's get a feeling for what Nature means by a very long time. You will be impressed by her patience.

At the short end of the scale lie species like the once-common limpet of the northwest Atlantic, *Lottia alveus*. If this species were still around we would probably have given it a user-friendly English name like eelgrass limpet, because eelgrass was all it ate. But it is not still around. In addition to food, it needed very salty water—the kind found in real ocean habitats rather than estuaries. After a new slime-mold disease wiped out the eelgrass from all marine waters in the early 1930s,[12] the limpet vanished utterly. All this happened within a few years.

The passenger pigeon took much longer to go. It fed on beech nuts and other forest seeds. But we needed those forests for farms. Beech tends to grow on good soil spread over a flat, easy-to-plow landscape. And beechwood isn't a bad timber product either! What an inviting combination for Americans of the eighteenth and nineteenth centuries. Worse still, passenger pigeons were good-sized birds and made tasty eating. They were harvested in great quantities by professional pigeoners. And so, we obliterated what had been an enormously successful species.

But it took us a while. A century or two, in fact. Martha, the last living passenger pigeon, fell dead from her perch at the Cincinnati Zoo in

September 1914, long after most of the continent's beech forest had been removed and a generation after the last pigeoners had given up the hunt.*

The Maori of New Zealand, with less technology, probably also took about a century or two to destroy the moas, leaving heaps of huge bird bones piled at old campfire sites. But it took nearly a whole millennium for some of the other extinctions they caused. Consider the case of the Stephen Island wren. Stephen Island is a lighthouse outpost in Cook Strait off New Zealand's South Island. In 1894, its lighthouse keeper got a cat that liked to prowl about the island looking for mischief. Over a period of some weeks, it returned with the bodies of a few odd-looking tiny birds. The keeper himself had seen them flitting about the island on his evening strolls, but he never thought much of them until the cat began hunting them down and hauling their remains back home. Then he realized that he had not seen quite their ilk ever before.

And he never would again. The lighthouse keeper's cat killed them all. But scientists studied their corpses and confirmed that these birds belonged to a newly discovered species. They dutifully gave it a Latin name, *Traversia lyalli*, and never saw a single one alive.

The Stephen Island wren was a member of a special bird family restricted to New Zealand. Not true wrens at all, the family consisted of five tiny-bodied species of wren-like songbirds. Now there are only two or three species left. The third is so rare that it too may be heading for oblivion. And the only species of any particular abundance, the rifleman, lives mostly in shrinking native forests. So we may lose the entire family. We sometimes keep separate track of such major losses and reckon them even sadder than the loss of a single species.

For decades, science used the extinction of the Stephen Island wren as an example of extinction caused by a predator (that cat, of course). But in the late 1980s, we learned that the cat had been merely the whiff of wind that blew away a species doomed for nearly a thousand years.[13] Subfossil bones revealed that the species had been abundant and widespread on mainland New Zealand when the Maori people colonized. We do not know exactly why the wrens lost their foothold in New Zealand

---

*Those pigeoners must have been a particularly relentless bunch. In the early 1880s, when pigeons were quite scarce and already doomed, parties of professionals were still tracking their whereabouts with the latest technology—the telegraph.

proper. Maybe it was the Polynesian rats imported by the Maori. Or maybe they depended upon some special habitat that the Maori also found useful. Or maybe the Maori had nothing to do with it. What matters here is the entire millennium it took them to become extinct. We must never underestimate the patience of extinction. If the cat had not dispatched them all, some storm or other catastrophe would have. A population of a few dozen little birds is a macabre joke; everyone knows such a species is terminally ill. Some scientists have even taken to calling them "zombie species"—the living dead.

We shall see that many species are now on the long road to extinction. For them, the handwriting is on the wall: 𝔐ene, mene, tekel and pharsin;* their days are numbered. But they will not disappear tomorrow.

How significant is it that extinction can take a very long time? Very significant if you can manage to get your torrential rains over with in 40 days and 40 nights. If that ends the trouble, you won't lose much more than the unicorn. But if your environmental insult lasts a long time, then eventually extinction will catch up and frustrate your most skilled and energetic efforts at ark building.

What kind of environmental insult is *Homo sapiens*? A 40-day quickie? Not likely. Unlike the floodwaters, we will not recede. We have no plans to vacate what we have taken—not in the next 40 days, not in the next 40 years. Not ever. So we have created an entirely different problem from Noah's. To survive in our shrunken world, natural species cannot merely play the role of passenger. They must participate fully in their voyage to the future. They must have sustainable populations, replenishing themselves generation after generation.

So, 40 years, let alone 40 days, won't do. We had best look far down the road. What do we foresee? How many species are heading for oblivion?

An Earth reduced to 5 percent of its size becomes a set of miniature continents like New Zealand or Hawaii. In chapter 8, we already looked at the sustainable diversity of such continents. They were depressed in proportion to the smallness of their areas. A continent with 30 percent of the area of South America, say, will have about 30 percent of her species diversity. A continent with 10 percent of her area will have about 10

Daniel 5:25. Bible scholars interpret these words of mystery to mean that doom is near.

percent of her species diversity. So things do not look too good for our shrunken Earth. With only 5 percent of her area left, she will be able to sustain only 5 percent of her species.

The ultimate damage, the 95 percent loss of species will happen because of the imbalance we have created in Nature's give and take, the imbalance in speciation and extinction rates.

When we reduce natural area, we increase extinction rates. The holes in our metaphorical leaky bucket get larger and more numerous, increasing the rate at which the bucket loses water. We can predict that the equilibrium water level will drop—nature will harbor fewer species.

But what we do to speciation is equally serious. Reduced area curtails speciation rates. It cuts the flow of water going into the bucket. That flow becomes a mini-trickle, decimating the power of Nature to restore the accelerated losses she is suffering.

How can it be otherwise? In a natural region, only three types of processes add to diversity, and two of these can provide no help at all to our shrunken world:

- The almost regular free-flow of individuals from source habitats to sink habitats is like a natural export-import business. But losing 95 percent of area puts most of the source habitats straight out of the export business. All the source habitats in the 95 percent of the continent that we expropriate no longer export any individuals. They aren't around.

- Island-style immigration of species from other regions can add to diversity. But our new mini-continent—unlike an island—cannot rely on immigrants to restore its extinct species. With difficulty, occasional individuals do manage to immigrate over space to islands. But they cannot colonize through time.

- Speciation. This is all that remains to sustain global diversity. Moreover, because of the reduced amount of area remaining to Nature, even this rate is diminished considerably.

And there we are. The residue of our natural world is a time island, not a space island. It could never by itself have produced the diversities it has today; it is far too small. Where did it get them? From a larger world that existed centuries ago. But those continents of centuries past are totally cut off from us now. They can contribute nothing else to our world. We are on our own. What we squander, we will lose forever. There will be no immigration rate to supplement whatever meager speciation rate we still have. No species that becomes extinct today can immigrate from the past to restore itself to the future.

This monumental loss of species will not happen instantly. Far from it. Such a vast extinction will take a vast amount of time. How long? Perhaps tens of thousands of years. But we won't lose species at a steady rate either. Most of our losses will come quickly in a first stage that will be a few human generations long. Then a prolonged second stage will begin. During the second stage, the rest of the losses will accumulate in tiny steps.

The First-Stage Loss

As our time island drifts farther and farther away from the year 1000, we will first lose the species whose source habitats have disappeared. How long will that take? It depends on how far below the break-even point a species finds itself. If its births almost match its deaths, then the species will take a long time to vanish. But if its death rate dwarfs its birth rate, then it will go quickly.

Because an island is like an area without sink species, we can predict the extent of the first-stage loss. We use our species-area rule.

Let's imagine ourselves alive in the year 1000. We measure the diversity of an entire continent and a 5 percent piece of it. The slope of the line connecting these two measurements in logarithmic space will be about 0.15.

Now we climb into our time machine and go forward to the time when all the sink species of today have gone extinct. If you need to imagine a year for that time, imagine it comes about a century from now: the year 2100. We need not select a 5 percent piece to census because that's all we have left; so we census all the remaining species on the continent. We put our result on the same piece of graph paper we used in the year 1000, and connect the old continent's point to that of the shrunken, 5 percent continent. The slope will be about 0.3. Why? Because islands have such slopes, and islands differ from pieces of mainland because islands lack sink species.

Now a simple formula gives us our prediction. Let's set the area and diversity of the old continent to 100 megablocks and 100 species so that our predictions emerge as percentages. Then the formula for the diversity of any smaller area is

$$\log S = 2 - 1.3z$$

where z is the slope. If we set z to 0.3, then $\log S$ is 1.61 and S is 40.74 species. That means only about 41 percent of the species will remain. Almost 59 percent are sink species and should disappear in the first stage of the adjustment to a mini-Earth.

Suppose we have the slope wrong? Suppose we have the area of mini-Earth wrong? In neither case does it matter very much. In the table, you see the result of leaving as little as 1 percent or as much as 10 percent of the Earth's land in a natural state. I combined these with two divergent estimates of the slope: 0.25 and 0.45. All four combinations spell rapid trouble for diversity.

Table of percentage extinctions caused by the disappearance of sink species
(Middle column assumes 1 percent natural land remains; right column assumes 10 percent)

Slope	% Extinctions	
0.25	68.4	43.8
0.45	87.4	64.5

A number of places in the world actually show us that we have the right prediction for the first stage of extinctions. As an example, consider the wheatbelt of Western Australia. This ecosystem stretches along a semiarid band between the inland desert and the coastal forests and heathlands of the state. If you clear it and plow it up, it makes superb wheat farms. And that, precisely, has been the fate of almost all of it during the past century.

But natural ecological communities do remain in a small fraction of the wheatbelt region. And the people of Western Australia have created a system of 22 isolated reserves to save the natural ecosystem of the wheatbelt. The reserves range in size from 34 to 5,119 hectares.[14]

Ecologists keep close track of many sorts of species in the wheatbelt reserves, but we should pay particular attention to two: lizards and birds.[15] Birds, you see, can really move around. So, for birds, the reserves are not so isolated from each other. We can expect sink species of birds in the reserves. Lizards on the other hand should have more trouble moving between reserves. So we expect each reserve to lose its lizard sink species. Thus if we look at the effect of reserve area on bird diversity, we should see a slope typical of mainland patches. But if we look at lizard diversity, we should see a slope typical of islands.

And so it is. The slope for birds is 0.17, an absolutely typical slope for birds in mainland patches. But the slope for lizards is 0.26, and that is what we expect from islands. It does indeed seem that each reserve has lost its sink species of lizards.

In the case of the lizards, we actually have diversities from a real set of 30 Western Australia islands. Its species-area slope is 0.33, a value typical of archipelagos, although a bit higher than the slope of the wheatbelt reserves. That suggests, unfortunately, that the wheatbelt reserves may still have a few lizards to lose. But the main point does not change: Isolate a place sufficiently and it loses its sink species.

The Second-Stage Loss

Once the sink species have become extinct, extinction rates plummet. Previously, there would have been many species whose populations declined relentlessly, year after year, until they had vanished entirely. However, once these are gone and stage two begins, the average reproductive rates of the remaining species are enough to replace their species successfully generation after generation.

Because all the remaining species will have reproductively healthy populations, one might think they would be safe. But the ecologist does not smile; the ecologist knows about accidents.

Accidents happen. There are good years and bad ones. Storms and earthquakes. El Niños and global change. Ice ages and greenhouse events. New forms of disease and parasites to resist. The Earth just won't hold still. Accidents at all scales threaten the existence of each and every species regardless of how well-adapted it may be. Experience teaches us that even extreme abundance is no guarantee of eternal existence. There is a time to live and a time to be a fossil.

And so Nature, our heroine, begins stage two of her tragedy, extinction of the successful species. Gradually, as accident after accident overtakes the remaining species and speciation lacks adequate rates to replace them, the mini-Earth's diversity drops lower and lower.

Of course, no one has lived long enough to observe a case of withering diversity, but careful observation of what is around today confirms our suspicions. The most transparent example comes from 17 islands in the Sea of Cortez,[16] ranging in area from 1 to 1,196 square kilometers.

As the glaciers melted after the last ice age, they added their water to the Earth's seas. Gradually, islands began to be cut off from mainland Mexico. Geologists can estimate the time at which these islands in the Sea of Cortez became isolated. The islands vary in age from 5,800 to 12,000 years.

Because birds immigrate well compared to other life-forms, the islands are bird islands in every sense of the term, geographical and biological. But lizards are not so mobile, as we have already seen in the wheatbelt of Western Australia. Lizard waifs, perhaps set adrift by some accident onto the Sea of Cortez, only very rarely get to islands before they starve or drown. So we can expect a quick decline in lizard species as sink species disappear on newly formed islands, followed by a much more gradual decline as accidents happen to source species.

The gradual decline leaves a trail of evidence that we can see by studying the lizard diversity on islands of different age. Sure enough, once we correct for the different areas of the islands and their different latitudes, older islands have fewer species than younger ones. An island of average area (10 square kilometers) and latitude (27° North) lost about four lizard species from the time it was 5,800 years old to the time it was 12,000 years old. That may not seem like much until you realize that at 5,800 years, it had only six or seven species meaning that such an island actually lost about 60 percent of its diversity in 6,200 years! And remember, the 60 percent figure may not be the end. We do not know how long Nature will take to finish the decline. If we had islands older than 12,000 years, they might very well have even fewer species.

And if we had islands that were even younger than 5,800 years, we would almost surely detect a level of extinction even higher than 60 percent. The 60 percent loss is only what happened *after* the initial 5,800 years. But a 10-square-kilometer patch of Mexican mainland has a lot more than seven species. Many of these disappeared during the first 5,800 years of island existence and we have no islands young enough to measure how many there were.

Think about it. The youngest of these islands is already 5,800 years old and yet the extinction trail is fresh. That reemphasizes how very long the extinction process can take. If 5,800 years on a small island is not long enough to complete it, we cannot pay much attention to the Pollyannas who tell us not to worry so much. Just because we have not yet paid the whole laundry bill for putting the Earth in hot water does not mean it won't come due! It will. Every study of the relationship of area to diversity tells us that. We and our children's children will pay it out in prolonged installments.

Although, in the fullness of time, our shrunken world will gradually lose more and more species, she is not going to lose everything. She will still be very large, even when nature occupies only 5 percent of the Earth's area. So every once-in-a-while, a new species will evolve. Eventually, there will be so few species left to suffer accidents that the rate of extinction will be no more than the trickle of speciation. The mini-Earth will rest at a new sustainable diversity. It will rest on an intercontinental line like the one connecting New Zealand and Oceania (in chapter 8).

Because that line's slope is approximately 1, we can predict this new diversity. It will be very nearly the same as the proportion of the original natural area that remains. So, for example, with the mini-Earth at 5 percent, diversity will also be at 5 percent. In other words, we stand to lose 95 percent of all species diversity. There will be no bargains. We'll get what we pay for.

In addition to all the area patterns that support our predictions, the fossil record actually records the history of an experiment that Nature herself accomplished over the past quarter of a billion years. It seems that Nature likes to play the accordion with the surface of the Earth. At times she floods major fractions of the continents with inland seas and the land area of the Earth shrinks. Then she dries the waters up and the land grows. Shrinks and grows. Shrinks and grows. Does diversity follow the music? Yes indeed.

Paleobiologists studying plant fossils in the northern hemisphere have discovered the dance. Using geological techniques, they estimate the positions and extents of the world's land surfaces at various times in the past. Then they count up the plant fossils from each period to see whether area mattered to diversity. It does, and the pattern is linear.[17] That means that the slope in log-space has a value of 1—just what we would expect. And just what we base our predictions on.

Thus, the natural world seems destined for trouble—long-term, very serious trouble. In fact, "an island adrift in a sea of time" does not quite describe Nature's current predicament. There's no drifting here at all. It's more like a free fall. Nature is falling into a black hole. As all sci-fi fans know, when falling into a black hole, you pass two event horizons. And so it is with Nature. Nature reaches the first event horizon when all its sink species are gone—when it has as many species as it would have if it were an island and the world of a thousand years ago were its continent. It reaches the second when finally it descends to its new sustainable diversity—when speciation and extinction rates again balance each other because only 5 percent of biodiversity remains.

Isaac Newton recognized the law of gravity. But he never understood it. He never could explain it. Neither could Albert Einstein. Even today, neither can we.

Still, all of us believe in gravity. None of us is ready to take a leap off the Eiffel Tower in hopes that gravity is merely a weak empirical generalization and not a law of the universe. So it is with the laws by which available area governs diversity. Surely we ought to improve our scientific understanding of them. However, we have the evidence of their power all over the world—and for hundreds of millions of years! In fact, we biologists are probably nearer their ultimate explanation than physicists are to an explanation of gravity. So why should we operate on the hope that area only *seems* to matter to biodiversity? Diversity soars on the broad wings of area. How can we expect it not to plummet once we clip those wings?

141

CHAPTER 10

Fighting for Crumbs: The Traditional Forms of Biological Conservation

A scientist must . . . be absolutely like a child.
If he sees a thing, he must say that he sees it,
whether it was what he thought he
was going to see or not . . .

Douglas Adams[1]

Functioning by themselves, the traditional forms of conservation are running out of time. Although they sure feel good—and right—in the end, conservation as we now practice it will only delay the monumental forces humanity has deployed against the world's diversity. Increasingly, traditional forms of conservation are becoming little more than diversions that occupy the attention of concerned citizens while the real struggle goes largely ignored.

I do not like having to write those words. I have been a conservationist for most of my life. I still proudly belong to conservation organizations like The Nature Conservancy and the Audubon Society—both of them leaders in reconciliation experiments. May I always belong to them. Such organizations educate us and mobilize us. They are effective advocates for the causes of Nature in the councils of power.

The organizations are not at fault. They want to save diversity. They are devoted to it. The fault lies with the particular strategy for saving diversity that dominates their agendas. And where do they get that strategy? From ecologists like me. So, it's my fault. *Mea culpa.*

Nevertheless, I write this chapter with considerable fear—fear that its words will be taken out of context, abused, misquoted, and sharpened to attack the cause I hold so dear. Yet I must tell the truth, particularly because the truth can lead us out of this mess.

Please permit me to immunize you against some of the misquotation. Simply reread the first sentence of this chapter. Notice the words **"by themselves."** Although the traditional forms of conservation by themselves cannot succeed, we must not abandon them. They will be an integral part of a strategy that can work.

Recall that the traditional forms of biological conservation are reservation ecology and restoration ecology. The two R's.

Reservation ecology says, We like this coral reef, we like this redwood stand, we like this forest of giant cacti. To preserve them, we will reserve them. We will not let ourselves be tempted to subvert them for profit. We will not exploit them. On behalf of the Earth, we will save them from ourselves.

Restoration ecology is more aggressive. It says, We are sorry to have fouled up this patch of the world. We wish we could have seen it when it was natural. We will determine what it would have been like way back when. And then we will make it like that again. We will restore it to Nature. We will bestow it upon the Earth and upon our children's children.

The two R's both resound with the highest human motives. Stewardship is moral. Nature can be beautiful and uplifting. There is nothing mean or greedy about relinquishing some of our power to exploit and expropriate. So what's wrong?

What's wrong is that traditional biological conservation rarely incorporates a dynamic view of Nature. It rarely asks what forces have brought a patch to its present state, the state we want to maintain. Not asking that question, it then fails to appreciate what Heraclitus taught us so long ago: Things change.

Just think about the effects of global warming on the current system of reserves. Each of these depends for its ecological character on the integrity of its climate. When that climate changes, the reserve topples;

its climatic rug gets yanked out from under it and we can no longer expect it to do its job.[2]

Then what happens? We can sell that nonreserve to exploiters, sure. But what will we buy to replace it? Some town or farm valley to which the right climate appears to have moved? Some shopping mall complete with a parking lot that we think looks like a good prospect to house the new reserve? And where will all the species take shelter while we take a few decades or centuries to make these new reserves competent to do the jobs of the old ones? Do we call in a cryogenicist and suspend their animation? How do you freeze an ecosystem?

I have seen a study of the effects of global warming on the reserves of Australia and it is frightening. There is little enough reserved already, but a degree or two of global warming and most of the appropriate climates get shifted out to sea where no amount of restoration ecology will help Australia get her reserves back.

In the past when climates changed, many species could track them. They could shift their ranges to maintain themselves in an environment that could support them. Paleobiologists looking at the effects of the ice age and its end about 10,000 years ago have followed the fates of trees and beetles as they scrambled to keep up with migrating isotherms. Most survived. But now where can they scramble to? Global climate change eliminates reserves, it doesn't just move them!

Even without global warming, patches of Nature will change. What you preserve today will be different tomorrow. In part, it will be different merely because you have reserved it. It will be different because you have treated it differently from its fellows. It will be different because it has become an isolate.

We have learned that isolates cannot maintain their diversities. Isolates are islands. Islands quickly lose their sink species.

But the situation may be even worse. Islands surrounded by water have an important advantage compared to reserves. Reserves are surrounded not by water but by novel terrestrial habitats that we have made. A few species, like house sparrows and starlings, use these habitats and become quite numerous on account of them. Those large populations of species that get along with us can send a stream of individuals into the edges of our reserves. There they can establish sink populations that deplete the resources we intended for the reserved species. If the reserve is small, it will have a large proportion of edge. Thus it is even possible that some of the source species in small reserves will lose their ability to survive in them. I wish this were just conjecture, but it is not. A famous experiment in Brazil set up reserves of different areas and demonstrated the pernicious effect of edge on the traditional inhabitants of the reserves.

So, you conclude, let's set up large reserves, and let's choose them very carefully in order that they preserve a complete set of natural habitats. A good idea, actually. We should do it and we are doing it. It will greatly slow the losses. But it too will fail in the long run.

First, it does nothing to prevent the disappearance of the reserves on account of climate change. Second, it cannot prevent the accidents that will gradually eat away at the set of species in the whole species pool—the pool, you will recall, of a thousand years ago from which those island reserves have drawn their diversities. If the sum of the areas of all the world's reserves amounts to only 5 percent of her original land area, the species pool itself will dwindle over a long period. And it will keep dwindling until, at 5 percent of its pristine diversity, it is small enough to be self-sustaining. We know, for example, that the magnificent diversities of the American tropics evolved and have been dynamically maintained in large measure because the American tropics covered a vast area. What if we save a few hectares? They will fall to a new and much lower dynamic balance in accord with the newly miniaturized status of their continent.

In other words, the area of a continental ecosystem is an integral part of that ecosystem. The area of an ecosystem strongly influences its extinction and speciation rates. These rates determine the number of species. As the final step in the sequence, virtually as an afterthought, natural selection sets about transforming the species so that they recognize many different habitats in the landscape. At first, we can save all the habitats that our current large diversities recognize. But over the long haul, the number of species determines the number of habitats. As we lose species, we will lose habitats too. That is why all of an ecosystem's area is needed to maintain its integrity.

So you cannot save a piece of an ecosystem. At least, you cannot save a piece of it for more than the short time it takes for Nature to start bringing this piece back into balance. If you do a good job and have a lot of good fortune, your piece will soon become a virtual reserve. It may look like a tropical forest. It may function like a tropical forest. It may even be a tropical forest. But it won't be the same as the forest you hoped to save. It won't be able to hold the wealth that it had when it was first reserved.

Why have you not heard this before? Two reasons.

First, the biology on which it rests—that continental area largely determines diversity—is fairly new. Ecologists made this discovery in the mid-1990s. It always takes a while for applied science to absorb new basic research.

Second, traditional conservation biology has emphasized trying to save individual species. That is why the chief legal weapon of conservation biology in the United States is the Endangered Species Act. Pin your hopes on the charismatic megafauna. Campaign for the warm and the fuzzy and the rare. Only recently have we become sophisticated enough to appreciate the possibility of saving an ecosystem.[3] And now that implacable tyrant, space, comes to teach us that the workings of an ecosystem have developed from its *entire area*, and cannot survive the curtailing of that area to a remnant.

When you want to save a species, you must of course stress saving its habitat. You work hard to find out what that habitat is and then you try to protect it. But you also study the genetic problems of rare species. And you want to know about their demographic problems too. How does chopping up the habitat into isolated bits further burden the species? How can we provide corridors of secondary habitat for our species to use—corridors that will fuse the bits of primary habitat into a supportive whole?

All such traditional activities of conservationists have considerable value. We shall need them no matter how our conservation strategy changes. But they rest on a single premise: They all hope to reduce—one species at a time—the high extinction rate that humanity has imposed on Nature.

Did you know that there are about 10 million—maybe even 50 million—animal species alive today on Earth?

Alex Trebek: "The USA spends about $350 million annually to help just one of these species in just one of its locations."

Jeopardy contestant: "What is the salmon of the Columbia River?"

To help these fish navigate their river, to get them past the massive concrete barriers that we need for flood control and power, the Army Corps of Engineers has been forced to build special barges. The barges serve as cruise ships for the lucky fish that are chosen to ride them.

I say! What a valiant and commendable effort! Long may it succeed. But it does not multiply. For a moment, let's ignore all the millions of insects and thousands of fishes and just do some multiplication with land-based birds and mammals. You know, the warm and the fuzzy. They are the tip of diversity's iceberg. How many species are there? Only about 20,000. At $350 million per species, that's $7,000,000,000,000. Per year. Ready for that, citizen-taxpayer? Maybe not? Hey, what's seven trillion between friends? You know, we ought to call it a gagabuck. One gagabuck doesn't sound like real money.

Now let's add some butterflies to enchant us and some bees to polli-nate our flowers. Oops, flowers! We forgot about flowers. Trees, too. There are quite a few thousand of those. This is going to cost more than the total annual income of the entire world. And the U.N. too.

It just does not multiply.

Of course, I know that I have erred. I've taken a single example and assumed it represents a typical case. I've also not allowed for economies of scale. Perhaps saving one species will make saving the next one easier and cheaper? But, I'm afraid that conserving species is a case where the economies of scale are negative. The bigger the job, the greater the cost per unit. Each extra species that we try to keep alive beyond the new, low, natural level of diversity in our shrunken world—each extra species— will cost us more, not less. Maybe I did oversimplify the multiplication, but I'll bet my mistakes caused underestimates in the species-by-species cost of diversity maintenance.

Two ecologists more optimistic than I have done a sober estimate of the cost of managing the rare and endangered species of the United States.[4] They examined the 681 listed species of plants and animals "imperiled to some extent by either alien species or fire suppression." These 681 constitute roughly one-third of all listed U.S. species. Wilcove and Chen estimated that to work on just the fire and alien problems of only these 681 species would take some $32 million to $42 million per year (1997 U.S. dollars)—some 90 percent of it for dealing with the alien species. They note, however, that in 1996 the U.S. Fish and Wildlife Service had only $36.5 million to spend on endangered species,* and much of that money goes for court battles, paperwork, and other nonmanagement expenses. They also believe that their estimates are "surely a significant

---

*The lucky salmon draw their funds from another U.S. pocket: the Army Corps of Engineers.

underestimate" because we do not know even what to do about many of the alien species, let alone how much it will cost. I might add (sadly) that, according to the authors themselves, they also underestimate the losses by ignoring the principal problem of most threatened species: loss of habitat.

I do not know a single conservation biologist who believes that our current strategies will avert a mass extinction. Most are trying, glumly and valiantly, to save what they can. A few have even relinquished that hope and are trying merely to collect, name, and document what we will soon lose. Even they have a long way to go. In three centuries, biology has classified about one or two million species. So we have not even halfway finished the process of counting the dead and dying. *The Red Book of Rare and Endangered Species*[5] is merely the Schindler's list of conservation biology—millions face obliteration, but it can defend only a few.

Nevertheless, we can do important things to reduce the rate of extinction in our reserves. We can. And we are. And we should. If we lose the whooping crane in a century or two because of some new disease or some freak storm that plows into coastal Texas during the crane's annual wintertime visit there, that will not please me a whit. But the whooper was headed for extinction in 1950. So our efforts will not have been wasted. We will have indeed reduced the rate of extinction. And maybe we won't lose the whooping crane for 100,000 years. Who knows when accidents will happen?

And who can complain about the heroic effort to restore at least a semblance of Florida's Everglades to ecological health? I find it truly thrilling to know that half of it may rise up from the functional dead because of the dedication and skill of a veritable army of ecologists. It is probably the largest, most ambitious restoration ecology project we will ever see.

But that's exactly the point. Here was the Everglades, an entire gem of an ecosystem, that was not too large, and not too settled, not too commercially valuable except as a tourist attraction, and not too far gone to help. There won't be many others that pass such a Goldilocks test. A patch of tallgrass prairie here and there, but no one contemplates the ecological restoration of the state of Iowa. We need its farms.

True, some species will never get along with us. *Kulturmeider* to the core. So we cannot dispense with our reserves. We will find no other way to help inveterate *kulturmeider*. But the notion that reserves can save most of the world's species is a fiction. Reserves are simply too small. Too pitifully few. And they themselves are endangered by global climate change. Reservation and restoration ecology face, with romantic ingenuousness, the expanding needs of an expanding human population.

Should you be tempted to rely on reservation and restoration alone, just remember Ambrose Bierce's fabulous aeronaut of 1898:

> An Ingenious Man . . . had built a flying-machine. . . . At the appointed moment, . . . he turned on the power. The machine immediately broke through the massive substructure upon which it was builded, and sank out of sight into the earth, the aeronaut springing out barely in time to save himself.
> "Well," said he, "I have done enough to demonstrate the correctness of my details. The defects are merely basic and fundamental."[6]

The Endangered Species Act (ESA) has helped only a little.[7] Most endangered species have not recovered despite the protection of that law. The U.S. Fish and Wildlife Service provided the following figures for the ESA after it had existed for almost 20 years:

Table of Results — United States Endangered Species Act
(Percentages apply to the set of species whose status is known)

	1990	1992
Number of species listed	581	711
Improving	57 (12.2%)	69 (13.4%)
Steady	181 (38.7%)	201 (39.0%)
Still declining	219 (46.8%)	232 (45.0%)
Extinct	11 (2.4%)	14 (2.7%)
Unknown	113	195

Moreover, even the few recoveries of species that have taken place since the birth of ESA cannot always be traced to it. For instance, the infamous snail darter—the tiny fish that spawned a political backlash when it got in the way of a new Tennessee dam—was delisted only because scientists discovered a number of previously unknown populations. The sad truth is that ESA protection has led to the increase of very few rare and endangered species. Attwater's prairie chicken shows us an example of the frustration that all too often accompanies the ESA. This grouse used to live in nearly all the coastal prairies of Texas. By 1937, people had taken most of its habitat, and only about 8,000 birds remained.[8] Protected by law since 1967, and by the ESA since 1973, its

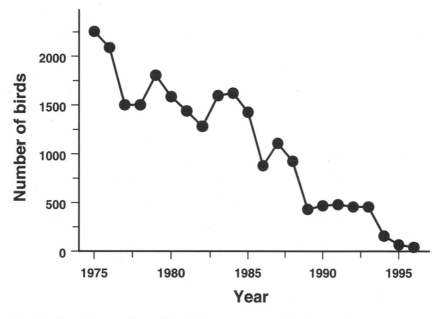

The decline of Attwater's prairie chicken — protected by the Endangered Species Act since 1973.

population continues to dwindle, as the chart shows.[9] The Endangered Species Act simply cannot help as much as we would hope, and anything we replace it with will do no better if it pretends that space is not all that important. Size matters.

It all reminds me of a story I once heard about the California legislature. It was the 1970s and geologists were working out why California so often visits earthquakes on its people. The legislators learned that the state is divided by the massive San Andreas Fault. The western part of California is slowly, inexorably traveling toward Alaska along that fault. Every time it slips another notch nor'northwest, the Earth trembles.

Eureka! The solution became obvious. Pin the suckers together! Yes, the same state that had gloriously rechanneled all that water from the Sierra Nevada to the farmers of its Central Valley, the same state that in so doing had dried up ancient Lake Tulare, the same state that had transformed the Los Angeles basin into a haven for sore eyes and emphysema, *that same state would pin the two sides of the San Andreas fault together.*

The way I heard it, the bill to establish the project almost passed. I cannot remember why it did not.

No matter. The story illustrates the delusional might of wishful thinking. The truth is that despite our powerful fantasies, we have almost no power to intervene successfully in the face of major environmental disasters. Maybe we can push our weight around and steal the real estate of every other species on the planet, but we are pretty puny when it comes to the big stuff. When we release enough chlorinated gas into the atmosphere to erode the ozone layer, we have to wait for Nature to clean it up. When we burn enough fuel to change the Earth's climate, we have to wait for Nature to restore it. Like an infant in diapers, we can make a mighty mess, but we have to wait for Mama to set things right.

And we certainly cannot add to the size of the Earth. Not to mention the gravitational disaster we would cause if we could! So, when we occupy land that previously supported millions of other species, Nature declares them losers and exterminates them. Those are just the rules of the game.

Vandermeer and Perfecto skewer the traditional strategy for the preservation of species diversity.[10] They begin by characterizing it most fairly: Set up reserves by buying unexploited land, then declare that land inviolable for other human uses. Thus they make us see that traditional conservation regards the preservation of biodiversity "as the antithesis of other types of human activities, such as forestry and agriculture." Then they go for the jugular: "The classical model thus strikes a Faustian bargain in which the inviolable lands are to be religiously respected and all other lands are to be ignored—let us have our 100 hectare wilderness and we will remain silent when the other 100,000 hectares of forest are destroyed."

We must abandon any expectation that reserves by themselves, whether pristine or restored, will do much more than collect crumbs. They are the 5 percent. We need to work on the 95 percent.

CHAPTER 11

Extinction Happens

Of all living creatures man
is the most dangerous—
to everything else that lives
as well as to himself.

Joseph Wood Krutch[1]

Nineteenth-century science laid most of the foundations of scientific understanding. It measured the speed of light. It discovered most of the universe's elements. It deduced the workings of natural selection. What a great time to be a scientist!

One of the monumental discoveries of nineteenth-century science lies virtually unsung in the subterranean foundations of twenty-first-century ecology. *Species become extinct.*

The discovery of extinction began in the late eighteenth century when scientists uncovered the bones and shells and leaves of life-forms that no one had ever seen alive. It was exciting stuff. Even Thomas Jefferson participated. He devoted an entire room of Monticello to his private collection of fossils, and he contributed whole boxes of specimens to the Peale Museum in Philadelphia, first in the nation to exhibit fossils scientifically.

But so what if fossils are real? That doesn't prove extinctions ever happened!

For a long time, the clergy argued with scientists about what fossils actually mean. Many scientists equated fossils with extinctions. But the clergy did not. Surely, maintained the holy men, fossils do not signify extinctions! How could it be, they wondered, that God—having put Noah to such trouble and having promised faithfully not to do it again—would tolerate the utter disappearance of any of One's own innocent creations? The dispute raged for decades.

The truth is that even the scientists quarreled among themselves. Despite the fossils, many doubted that extinction had ever befallen any species. Suppose—they worried reasonably—that fossils actually do represent the remains of species that no longer live where the fossils were found. Yet, perhaps these species survive somewhere else? How can we be so sure they are extinct?

You see, the Earth of that time was a little-known place to Westerners. Maybe, they thought, some remote tropical mountain concealed the mammoths and the dinosaurs. Maybe weird trilobites still patrolled some unexplored ocean floor in a remote sea. After all, fabulous discoveries of strange species were pouring in to Western capitals. Jefferson himself hoped for the discovery of living giant ground sloths and mastodons in the "unexplored" vastness of the Louisiana Purchase that he had bought from France on behalf of the United States. Who could be sure of any extinction? In Jefferson's words, "Such is the economy of nature that no instance can be produced of her having permitted any one race of her animals to become extinct."[2]

Eventually, though, all scientists had to admit the reality of extinction. Meriwether Lewis came back from his expedition with Clark and reported neither seeing nor hearing of a single "living" fossil. No one would ever again see a living ground sloth, an imperial mammoth or a sabre-tooth. Such disappointments were matched time after time in the scientific societies of the world. In Paris, in London, in Berlin, science had discovered extinction.

Today, the reality of extinction is widely acknowledged. Even the Pope has come to agree. Yet while our heads readily accept the concept of extinction, I'm afraid our hearts remain stone.

We have not begun to accept extinction on an emotional level. In our gut, we are sure it never really happens: No, no, no. It cannot be! Like death, extinction is something that someone else suffers. It happens to someone else's animals. It snuffs out flowers on someone else's planet. *Our* world is immortal.

This a monumental problem. As Dale Dauten has put it, "... the truth is much more than the facts. You know the facts, but feel the truth."[3] If we do not feel the truth of extinction—if we continue to expect the impossible—we can never commit ourselves to the effort of reconciliation ecology. Feeling the truth of extinction must precede the effort to thwart it.

Many stories of human improvidence lead to the conclusion that most people do not yet accept the truth of extinction. Let me tell you a few. I'll begin by introducing you to David Steadman.

Dave Steadman took his Ph.D. from Arizona in 1987. He is a combination anthropologist and biologist. Dave studies the epic extinctions produced by Melanesian, Micronesian, and Polynesian societies wherever they colonized in the Pacific.[4]

Many centuries ago, the Pacific islanders hopped onto their great seagoing, double-hulled canoes with their pigs, their chickens, their dogs, and their little Polynesian rats. They took along their taro roots, yams, bananas, sugarcane, and barkcloth trees, and they spread out from their southeast Asian homeland. They got as far west as Madagascar, and traveled east to Hawaii, Easter Island, Tahiti, New Zealand, and to all the lovely islands in between. Wherever they went, they made big ecological trouble. Every island that they colonized lost a huge proportion of its plant and animal species. Dave Steadman has shown us that.

Does it surprise you that non-European peoples also cause extinctions? It is fantasy that only Europeans make ecological trouble. It is fiction that the other peoples of the world live in perfect harmony with nature. People are people. They just vary a little bit in culture and technology from time to time and from place to place.

Did the Pacific islanders intend any of the extinctions? Probably not. Most likely all of them were accidental extinctions: for example, all the great and little moas (11 different species) whose bones are found today near the ashes of long dead Maori cooking fires in New Zealand—they all vanished about a century after the Maori arrived (ca. 1250) and began to depend on them for food.[5]

Nor do I think that Kamehameha I (1758–1819), first king of the Hawaiian Islands, intended the extinction of the Hawaii *mamo*. A songbird about nine inches long, with a Jimmy Durante sickle of a nose for its beak, the Hawaii *mamo* belonged to the Hawaii honeycreeper family. Found only in Hawaii, this extraordinary family once comprised some

45 species. Their bills told the story of their broad evolutionary radiation. Some were conical, revealing a seed-eating habit. Some were shaped like insect eaters. Some had two shapes—one for the upper bill and a second for the lower. And quite a few, like the *mamos*, had the curved bill of a nectar feeder. Of the 45 species, at least 25 are extinct and many others are on the ropes. Perhaps as few as five can be considered safe.

A number of the nectar eaters had brilliant red or yellow feathers. A special guild of workers, the *po'e kaha'i*, had the job of catching these birds so that the nobility of the islands could have capes and hats made of such feathers. The greater the person's rank, the larger and more costly his cape. It's an old story, everywhere.

A moa leading its chicks. A thousand years ago, there were eleven different moa species. None survived their encounter with Man. Drawn and copyright by Geoffrey Cox, Rotorua, New Zealand.

Although the Hawaii *mamo* was mostly black, it had a few golden yellow feathers in its rump and under its tail. To exhibit his wealth and power, Kamehameha ordered a huge cape made entirely of those feathers. "Make it so," we can imagine he said to some royal pooh-bah. As a result, an estimated 80,000 *mamos* were killed to make a cape six feet long and eight to nine feet wide. The species never recovered, the very last individual being seen in 1898, when my own grandfather was a young man of nineteen.

A Hawaii mamo (R.I.P. circa 1898). Illustration by D.M. Reid-Henry. Most feathers that appear white in this illustration were actually golden yellow.

I have never had the heart to visit the splendor of Kamehameha's cape in Honolulu's Bishop Museum. Unlike the *mamo* itself, it survives—a major tourist attraction. Still, Kamehameha probably held the *mamo* in high esteem. He might even have executed any of his subjects who planned its extinction. But because he simply could not imagine that extinction, he became responsible for it himself.

Here lies the Hawaii *mamo*, done to death by a vainglorious, petty Polynesian potentate.

Of course, no one can ever prove that Kamehameha's cape resulted in the extermination of the *mamo*. In fact, it took some eight or nine human generations to complete, so the 80,000 Hawaii *mamos* that "donated" to it, were spread out over many bird generations. And some claim that the bird snares used by the *po'e kaha'i* were live-traps. Hence, they maintain, most *mamo* individuals survived their unanesthetized plucking (having suffered only terror, pain, and indignity). They point to diseases like avian malaria as the true *coup de grâce*. Are these claims merely a politically correct protest against any hint that Polynesians too may lack a degree of ecological sensitivity and responsibility?

Maybe so. But maybe such claims are more than that. Maybe avian malaria really was responsible. We shall never know for certain because there is no way to bring the *mamo* back and try a different set of Royal Hawaiian policies.

But even if the details of my story of the *mamo*'s extinction are allegorical, their substance is true. Again and again, mass extinctions followed swiftly upon the heels of Polynesian colonization in one after another Pacific archipelago. Hawaii was the rule, not the exception.

Consigned to oblivion with the *mamo* are all four native tree species of Easter Island, the most remote outpost of Polynesia. Easter Island, a 65-square-mile patch of isolation, today called Rapa Nui by its residents, was first settled about 1,200 to 1,300 years ago. At the time, forests of a large palm called *niu* and trees of the basswood family called *hau* grew on its soil. The first settlers must have been glad to find the trees because Easter Island gets cool in the winter. Storms blowing up from the Southern Ocean sweep its shores and carry with them a tiny taste of Antarctic chill. Burning the wood from those trees surely made for a cozier winter life on Easter Island. Moreover, early islanders got more than heat from their trees.

Polynesians are a nautical people. They travel widely over the ocean and they fish in it, too. The first Easter Islanders were no exceptions.[6] Their centuries-old accumulations of garbage reveal that major parts of their diet—and virtually all their protein—came from fish. Fishing takes boats. Boats require lumber. Lumber comes from trees. The new residents of Easter Island used the trunks of the *niu* and the rope they made from the *hau* to build oceangoing canoes to fish in far-off places. Evidence points to destinations as far away as the islet of Sala-y-Gomez, some 250 miles to the east-northeast. Mt. Terevaka, which once harbored a forest of hardwood trees, means "to pull out canoes."[7]

But the trees suffered terribly. The pollen found buried in lake sediment on Easter Island reveals the unmistakable record of an ecological disaster.[8] The larger trees began a precipitous decline just about the time that Polynesians first settled the island. Within four centuries, the trees were nearly gone. About 1960, the very last of the trees growing on the island, the last individual *toromiro*, died.*

On Easter Day, 1722, perhaps 1,000 years after the first Polynesians had landed, Easter Island welcomed its first European visitors.[9] An expedition led by the intrepid Dutch explorer Jacob Roggeveen had come to call. But no great dugout canoes rowed out to greet it. Instead, most of the curious swam out, the ships being anchored about two miles offshore. And many did come in small canoes. Roggeveen describes them as poor and flimsy craft, fitted together from a number of narrow boards and leaking so badly that their occupants had to spend half their time bailing. He tells us why. There was simply no heavy timber on the whole island!

In 1770, Easter Island got a second visit, this time from Don Felipe Gonzalez who had come to claim the island for the Spanish crown. This time, only two little canoes paddled out to the warship San Lorenzo. The journal of one of the officers discloses that the island had not one single tree capable of furnishing a plank so much as six inches wide. Captain James Cook confirmed these observations when he visited in 1774. So did the botanist Forster in 1777.

Captain Cook opined that the canoes of Easter Island were the least seaworthy in all the Pacific! In 1786, Admiral Jean de La Pérouse brought home a drawing of one of these fragilities. He saw only three and thought it probable that, owing to the lack of suitable wood, not one would remain after a short time. He was correct. But it was not a great loss. The real damage had already been done centuries before.

Their language and culture make clear to us that the Rapanui[†] did not intend to destroy all the timber of their island. *Rakau* means "timber" or

---

*Alone among the island's tree species, *Sophora toromiro* survives in a few botanical gardens around the world.

[†]Rapanui, the modern name for the Polynesian people, language, and culture of the island, was first applied to Easter Island by Tahitian sailors in the 1860s.

"tree branch" in most of Polynesia, but in Rapanui it means "wealth" or "treasure."[10] As a sign of status, Rapanui royalty wore *reimiro*—a wooden pectoral pendant in the shape of a canoe. But most telling may be the myth of Tangaroa on Easter Island.

A reimiro or canoe-shaped wooden pendant worn by Rapanui's kings and princes. Drawn by Greg Cooke, Arizona House of Graphics. © 2002, Evolutionary Ecology Ltd.

Among the Maori of New Zealand, and among the Marquesans, too, Tangaroa was the god of wind, sea, and fishing. In Samoa and Tonga, he was the god of craftsmen, first builder of houses and boats. But when Tangaroa came to Easter Island— so say the Rapanui themselves—the people killed him. They tied him up and clubbed him and he cried out, "Let me go! I am Tangaroa." But they killed him anyway. And they profited not. No one could figure out how to cook his flesh.[11]

The loss of their fishing fleet shows up as a profound change in the menu of the Rapanui.[12] Fish that can be caught only from seaworthy canoes do not occur in newer garbage heaps. Well, at least the fish were safe.

Asher Shockley, a prominent student of extinction (whose name I disguise, as I do the other details of the following true story), tells a tragicomic tale that unmasks the face of human improvidence. Shockley was working on Little Temple Island in the Ryukyus. A good friend invited him to go and see the last 26 pairs of Ryukyu petrels, a rare sea bird that nested on the coast of Little Temple. Once, it had been abundant and widespread, but now it was down to a struggling remnant. Shockley was happy to get the chance to see them. Who knew how much longer they could hold on?

Shockley's friend is a young, brilliant local official with a superb higher education from the University of Adelaide in Australia. Dedicated and selfless, he had returned home to serve his people. As Shockley and his friend drove north on that day, they chatted about the same things that you and I might if we had been along. The weather, the magnificence of their environment, good times they had shared. Soon, they arrived and got out of their car. The birds were there all right, close, and more or less on display.

The young official went back to the car to fetch . . . his shotgun! Ignoring Shockley's vehement protests, he shot all the petrels. "You Westerners do not understand," he declared. "They're a local delicacy. Very tasty. Why don't you come over for dinner and see for yourself?"

"But you just killed the last one," choked Shockley. "They're extinct now!"

"Nonsense, Asher. They've always been here. They'll be back."

I am not singling out Pacific Islanders for blame. Their problem is not being Pacific Islanders, it is being human. We are all of us to blame. European whites often deny that and lavish blame solely upon themselves.

It can hurt to look in the mirror. What is wrong with us? Why do we do such stupid things? Our world suffers. We suffer. We seem addicted to idiocy. We act like spoiled children with an endless supply of toys to break. What makes us so sure God will replace what we, so thoughtlessly and so cruelly, trample into eternity?

As Shakespeare put it, the trouble is not in the stars, but in ourselves. Science has been teaching us about extinctions for nearly two centuries. Our minds have learned their lessons well, but we fail to internalize the reality of extinction. We still feel it only as an abstraction, which is to say we don't feel it at all. Thus, we do not accept it as true.

Why do we deny the reality of extinction? Because natural selection never gave us a sense of the finite. That may turn out to be her fatal failure. As *Homo sapiens* evolved, natural selection put no premium on believing in limits to resources. She never paid us to imagine that we were about to spear the last living woolly rhinoceros, to boomerang the last of the giant kangaroos, to ensnare the last moa. On the contrary, human societies have almost always succeeded by imagining the world to be infinite, and encouraging their members to behave accordingly. Everything is possible. Can do! "We have a psyche predisposed to take from the environment with little thought for the future."[13]

So sorry! Extinction happens. Whole species keel over and die. Dodos and Carolina parakeets. Great auks and Stellar's sea cows. Animals that we loved, and many that we did not even know about. Flowers we cherished and those we never noticed. Trees whose majesty has inspired us. And weeds and pests and even germs. There is a time to live and a time to be a fossil.

Yet, if extinction is so natural, maybe there is nothing to do but let it happen. Let nature blow them all away like smoke. After all, well over 99 percent of the species that have lived are already extinct. If we know about them at all, it is only through their fossils. George Gaylord Simpson, my late colleague and one of the twentieth century's greatest paleontologists, called the Earth "a charnel house for species."[14]

In fact, even mass extinctions have happened before. From time to time, diverse catastrophes have smothered most of the Earth's diversity.[15] The dinosaurs and their friends were wiped out by the atmospheric consequences of a meteorite impact some 65 million years ago. About 67 percent of all species vanished. And that was not even the worst catastrophe life has suffered. About 250 million years ago, the trilobites and most of their associates checked out forever. Approximately 97 percent of all species died. Some believe a meteorite impact also caused this mass extinction, triggering the explosion of volcanoes worldwide and making it even worse than the one that extirpated the dinosaurs.[16] Others have found clues suggesting a major increase in the atmospheric concentration of carbon dioxide at the time. Enough excess carbon dioxide would have put them to sleep forever because carbon dioxide is a powerful anesthetic for many invertebrate animals. Yet whatever the cause, the Earth eventually healed herself.

The giant lemur of Madagascar (*Megaladapis*). About the size of a two-year-old steer, this species survived until almost "yesterday." It was driven to extinction, probably by humans, about three centuries ago. Drawing by Margaret Lambert Newman.

A terror crane (*Diatryma*) attacking a condylarth. Both these forms flourished in the Paleocene, some 60 million years ago. Man had nothing whatever to do with their extinction; they vanished tens of millions of years before we evolved. Drawing by Margaret Lambert Newman.

So, you may scold me, What a Chicken Little! We may be a little mischievous. We may even be trouble. But we're not that bad. Mother Earth always seems to recuperate somehow.

But Nature takes a very long time to recover its diversity after the cataclysm of a mass extinction. The fossil record tells us that recovery takes between one million and ten million years. A mere tick or two on the geological clock. But not on ours.

No one really intends to tell their great-grandchildren that, although they themselves will never see the riches of our planet, their descendants will . . . in a couple of million years. Furthermore, in this case we cannot be so positive that the world will ever recover, not even in ten million years.

The Earth's past tragedies were caused by temporary conditions. The dust kicked up by a meteorite, even if it hangs over the Earth for a decade or two, soon dissipates. That leaves the world to revitalize itself, to lick its wounds. To get on with life's processes, even as it buries its dead in the stony strata of the eons.

But the tribulations that humanity brings to the natural world may not be as temporary as the dust that follows a meteorite impact. Indeed, we plan them to be permanent. We intend to stick around for a very long time. If you had any doubt, consider the leap seconds that we add to our atomic clocks so as to compensate for variations in the Earth's rotation. Since 1972, the U.S. National Institute of Standards & Technology along with the U.S. Naval Observatory have added 22 leap seconds to our clocks so that, in 10,000 years, we will not experience dawn when the clock strikes noon.

If we do stick around for 10,000, or 100,000, or 1,000,000 years, how can the Earth ever heal? If we constantly reopen its wounds, how can the Earth ever mend? And how can it ever mend if we keep changing the kind of wounds we inflict?

You see, renewal of the Earth after a mass extinction depends upon a few well-known biological processes. The most general of these is natural selection itself. Dave Barry almost got it right when he wrote: "For example, if the climate was very hot, the animals without air conditioning died. If the climate had daytime television, the animals without small brains died."[17] Natural selection has the job of modifying species so that they overcome new environmental challenges. If nothing else, the works of civilization surely represent new environmental challenges!

But natural selection operates at a painfully slow pace. Its adjustments to the world of human-made habitats most probably cannot keep up with the rate at which we alter the challenges themselves. Indeed, it is a rare species that can move—like the house sparrow actually did—from a world in which its food comes from the undigested seeds left in horse manure, to one in which horses themselves have nearly disappeared from our lives. Natural selection is not so good at tracking a moving target.

Norman Myers, who first rubbed our noses into what we are doing to the planet's tropical rainforests, now wonders whether we are killing Darwin's genie (natural selection) at the same time as the species it fashioned. There is clearly no cause for optimism. So I would implore you, beg you—What the devil! I would even beseech you—to forget about natural recovery from mass extinction. It may never happen. And even if it did, it would take such a frightfully long time that for a hundred thousand generations, we would stand guilty of depriving the world of its riches. In the words of that great modern champion of diversity, E. O. Wilson, "This is the folly our descendants are least likely to forgive us."[18]

In short, I am not unaware of the history of epic natural disasters and the Earth's recovery from them. I just do not believe we need to allow one to happen on our watch.

We need to admit that there are rules and we are breaking them. Thousands of species that nature destined to last for hundreds of thousands or even millions of years may vanish in a few decades. Uncounted myriads will almost surely disappear in a few centuries. We are slashing at the fabric of our world and it is leaking species through its wounds, leaking much, much faster than nature can pour in replacements. The result promises to be a devastation, a horrible loss of diversity that no one really wants or intends. Accept that. Feel the truth. Once we feel the impending shock and, yes, even shed tears for diversity's losses (as I have seen my students do), then and only then can we get on with the serious business at hand—reconciling the habitats of our world so as to stop the catastrophe in its tracks.

CHAPTER 12

Clearing Hurdles

"Hope" is the thing with feathers—
That perches in the soul—

Emily Dickinson[1]

The science of the previous few chapters may make it seem that reconciliation ecology is inevitable. The examples of the first seven chapters may make it seem that reconciliation ecology has arrived. You may now believe that all a concerned citizen needs to do is applaud and participate. Not quite.

Reconciliation ecology lacks recognition and organization. We must discover how to integrate it with other conservation efforts. And it needs far more support. One person, most certainly including yours truly, cannot alone supply the necessary support.

Similarly, one book alone cannot supply all the answers. But it can raise some questions. That is this chapter's job. We start with the organizational issue, advance to integration, and finish with some support problems that actually turn out to be secret weapons for accomplishing reconciliation.

The Role of Government

Don't we need a federal agency to supervise reconciliation ecology? And don't they need to cooperate with a U.N. agency of some sort? Not at all.

I won't claim that governments have no role to play. Reconciliation ecology needs them. But just as surely, government needs to recognize that its role must remain limited and functional. Now is the time for work, not talking about work. Now is the time for the efficiencies of the smallest practical scale, not the power-hungry games of the empire builders. Now is the time for moral policies, not amoral posturing.

Thomas Jefferson said, "That government governs best which governs least." I understand that to mean that government governs best which does only those things people want done, but cannot do for themselves. And that government governs best which does those things at the smallest practical level. That's not as catchy as Jefferson's version, but it will help us to organize reconciliation ecology effectively.

The European Union captures the meat of the idea with one of its essential principles: It is called "subsidiarity." For the European Union, subsidiarity means "the Community shall take action only if the objectives of the proposed action cannot be sufficiently achieved by the Member States, and can therefore, by reason of the scale or effects of the proposed action be better achieved by the Community."[2] In less legal-sounding words, subsidiarity means that action should be taken only by the smallest level of human organization that can bring it off successfully.[3]

John McCormick terms subsidiarity "one of the defining mantras of the European Union."[4] But conservationists can adopt it, too. Admittedly an ugly word, subsidiarity is a beautiful idea. Small is beautiful. Small can be efficient. Small harnesses the immense power of the citizen-conservationist. Subsidiarity forever!

Consider the various examples of reconciliation in progress. If a free and voluntary neighborhood association can protect a prairie wildflower in a set of backyard habitats, why should they have to justify themselves to any level of government? As I see it, their sole responsibility would be to let others know what they are doing. That way, they can avoid unintended duplication of effort.

And if a private farmer sees that using reconciled methods, she can make more profit over a longer period, then the government's job should be to get out of her way. Same thing when a land developer begins to understand that the bulldozer often reduces his profits. Governments should figure out what they can do to create self-regulating systems that encourage reconciliation and sustainable, profitable yields. And then they should make themselves scarce.

As it now stands, governments do not make themselves scarce. In my opinion, they interfere most often at the urging of conservationists. We conservationists realize that governments do indeed seem to possess the necessary power to protect environments. So we often turn to our governments for laws to protect Nature. And sometimes, we do it effectively. We have found ways to approach governments with our agendas. But such laws—well intended always, and often useful, at least initially—may sometimes wear out their welcome. For one thing, they eventually cast us as Johnny-one-notes, making our campaign seem to others like thoughtless reflex actions.

The Endangered Species Act gives us a good example of the true complexities we face. What conservationist can imagine a world without some government protection of endangered species? What else but a set of governments can prohibit trade in things like ivory and rhino horn, sea otter pelts and egret feathers?

Yet some of the provisions of the Endangered Species Act have led us astray. We saw how the ESA's rules discourage habitat protection on private property. I will add that the Endangered Species Act often inhibits the scientific research needed to help rare species survive. It's not just the chilling effect of massive amounts of paperwork. It's not just the fear that a mistake, however honest, could lead to arrest and prosecution. It's that saving many of these species will entail real experiments, such as those involving habitat manipulation. Most governmental administrators of the act believe that we ought to restrict the science done on endangered species. Their understanding of the law allows us only to observe them and their habitats. Most often, that will amount to the meticulous recording of extinction after extinction.

And did you know that most U.S. state governments have laws that prohibit collecting the seeds of endangered plant species and planting them in new places on private property?[5] Fortunately, these laws are rarely enforced. Some commercial seed suppliers make their living ignoring them. They sell quantities of such seed, ensuring that these rarities get the chance to spread into many home gardens. Nevertheless, it is too bad that the laws exist at all. They actually prohibit some forms of reconciliation ecology.

Still, let's not descend into anarchy. We need government to get involved in reconciliation ecology. Very often its role will be crucial. For instance, as in the case of Eglin Air Force Base, the government itself holds title to the land. If it abstains from reconciliation ecology, none will get done on its extensive property. In the western United States, that is serious. In many western states, there is more government-run land than private land. In many countries, the government holds title to even a higher proportion of the land than in the western United States.

We also need governments when we want to protect an animal like a large carnivore. That job takes a lot of land. Your backyard won't do. Even if I could work out a plan to save cougars in a collection of small backyards, do you really want them roaming your neighborhood, perhaps picking off a jogger every once in a while? On the other hand, the amount of real estate in a great national park might be enough to save cougars without sacrificing any joggers. Hence, governments surely need to establish large nature reserves, like national parks, to allow the earth's larger species to survive.

Moreover, without governments and large-scale action, how can we deal with the many species that cross international borders? These species come in two varieties, migrants and exotics. We want to defend the migrants and exclude the exotics. What good will it do to take care of the Mexican wintering grounds of monarch butterflies if they have no place to breed in the United States? And what good will it do to eradicate goatheads from our personal backyards after careless government policies have allowed this Mediterranean pest to break out of its natural home to invade and re-invade our landscapes.

The case of migrant species is pretty obvious. International treaties on migratory birds have long recognized the need for governments to cooperate. Acting alone is a pure waste of time.

Regrettably, however, governments have not often appreciated the usefulness of banding together to stop the introduction of exotics. Time and again, exotic species become pests, the very accidents that can hasten the extinctions of many other species. European starlings and house sparrows. Red foxes, burros, brown trout and bullfrogs. Tamarisks and Chilean pepper trees. All these are desirable where they are at home, where they belong to the native species list. But they often burst out of control when introduced far away.

The source of the exotic invader does not seem to matter. Australia detests Europe's contribution of the red fox to the species list Down Under. Europe detests the contribution of the gray squirrel from North

America. And North America would be a lot better off without its host of Mediterranean weeds.

Some governments do recognize the threat of the exotic danger. They take action by themselves to prevent its getting worse. The United States forbids the importation of mongooses to the mainland, for example. (They already live in Hawaii and Puerto Rico.) And have you ever been in an arriving planeload of passengers waiting to get off in Hawaii while an agriculture agent sprays the cabin to kill unwanted insect hitchhikers? I don't know how effective this is, but at least it bows in the direction of responsibility.

Western Australia has the most astonishing story to tell. It is the single place I have ever visited that lacks both house sparrows and European starlings. Both these avian cockroaches live in central and eastern Australia. How have the Western Aussies kept them out?

To the north, Western Australia's hot desert defends her. The starlings and sparrows would fry if they tried to penetrate from that direction. So the nearest practical source of the birds is South Australia. South Australia and Western Australia are connected by a narrow strip of tolerable habitat that runs along the shore of the Southern Ocean. This strip, called the Nullarbor Plain, provides a corridor for eastern species and western species to interchange. But Western Australians are having none of it.

Western Australia stations a defense team on the Nullarbor Plain. The team searches out and catches the unwanted birds as they try to funnel through. So far it has worked. The parks and streets of Perth are filled with the parrots and honeyeaters that belong there.

Western Australia's methods may seem extreme, and they are. But so is the threat from a robust, exotic invader. Perhaps Tom Lehrer was wrong when he sang, "It's not against any religion to want to dispose of a pigeon."[6] But only slightly. Reconciliation ecology may make peace with and between all sorts of people. It may find room for all sorts of animals and plants. But it must zealously try to extirpate many nonnative pest species from its midst. That we cannot do without help from our governments.

The Republic of South Africa has provided a model of what can be done and how beneficial it can be. From chapter 8, you will remember the astonishing diversity of her fynbos plants: The shrublands of the Western Cape Province, a relatively tiny area of about 90,000 square kilometers at the southwestern tip of Africa, support approximately 8,500 plant

169

species. About 68 percent of them grow nowhere else. Many of our most beautiful and familiar flowers, such as geraniums and gladioluses are fynbos natives. But the wealth of the fynbos is in danger.

Over the centuries, we have introduced hundreds of exotic tree species into South Africa. Of these, about fifteen have caused serious trouble, invading the landscape and elbowing out the natives. Pines and mesquites from North America and Europe along with acacias, eucalypts, and hakeas from Australia, have taken over patch after patch, especially after the frequent fires that are such a prominent feature of fynbos ecosystems. In the more mountainous parts of the fynbos, invaders already covered some 31 percent of the land by the end of the twentieth century. At the rate they were taking over, they were expected to cover some 80 to 100 percent by the end of the twenty-first. You might think that the future of this bastion of plant diversity looks hopeless, but read on.

Exotic invaders threaten Western Cape Province with more than loss of its plant diversity. The Province depends on the watersheds of the mountain fynbos for two-thirds of its fresh water. Because the exotics are trees rather than shrubs, they are larger and use more water than the native shrubs. So far, they have reduced the water supply of the region almost 7 percent. Eventually, if the exotics are not controlled, water supply will decrease about 30 percent. Keep in mind, this province is semi-arid—it has no water to spare.

What is the usual response of a government to a water shortage in a severely seasonal environment like Western Cape Province? Bank more water. Build more dams and reservoirs. Moreover, South Africans need more jobs and the jobs from such construction would help. But the province followed a different path.

A savvy set of South African scientists[7] decided to study the economic impact of an alternative to dams and reservoirs. They reasoned that because exotic trees are the problem, perhaps the solution is as simple as hiring people to eradicate the exotic trees! They were on to something.

Building dams and managing a watershed would cost less than controlling the exotic trees. But the difference would be small. Clearing exotics would cost only 11 percent more than building dams. Meanwhile, the extra 11 percent would yield an immense amount of extra water, 29 percent more than the dams. All in all, the method of clearing would reduce the cost of each gallon of water by 14 percent and bring more of it to a province that had little prospect of getting more by any other means.

The government of South Africa had to try this. Because the plan involved hiring many people to help clear the exotic trees, it named the program "Working for Water." Working for Water succeeded well enough during a trial year that it received a handsome boost in government support to $50 million (U.S.) per year, enough to employ 35,000 people steadily.

What a twisted bit of good luck! South Africa's other problems—her need for jobs and water—may save the plant diversity of her mountain fynbos. Still, without the government to capitalize on that luck, it would have gone to waste. In the mountain fynbos, government planning and government action were required to reconcile human use of the land as watersheds with its occupation by the diverse assemblage of plant species that had been born there.

"Think globally; act locally."[8] It's René Dubos's great slogan. It's also a great place to begin. But conservation biology is a science, not a religion. And science is the most hard-headed of human pursuits. It relentlessly changes to accommodate the facts as they become known. So, despite its fame, we need to amend the Dubos slogan slightly.

"Think globally; act locally—unless you need to act on a larger scale to accomplish anything."

Actually, anyone who reads the Ward and Dubos impassioned but reasonable plea for global planning and global cooperation on behalf of the environment knows that René Dubos would approve that amendment without hesitation.[9]

Whither Yellowstone?—
The Role of Nature Reserves

In 1998, we celebrated a most unlikely event: the establishment, 125 years before, of Yellowstone National Park. Yellowstone was the world's first national park. I attended its birthday party—an orgy of learned speeches—at Montana State University. It lasted several days.

But what made Yellowstone possible? How could America have given birth to it in 1873? Think about it. *Homo sapiens* is a very conservative species. Any new idea will always have trouble getting past the drawing board stage. But this new idea? It was preposterous. Victorian Americans lived in the full bounty of the wildernesses of the West. To set aside a large wilderness tract inside a vast wilderness of unreckonable immensity? How profoundly silly. How insane. So Yellowstone was not simply a new idea. It was a radical innovation, an apparently mad, mindless, and useless caprice.

I am not making this up. Yellowstone was indeed a tough sell. The Congress of the United States had to be pestered for twenty years by visionaries, outside its walls and inside, too—and even by the likes of the Northern Pacific Railroad. (The Northern Pacific's visionaries saw tourists and money. Were they ever right!) And if you don't believe how tough it must have been to sell Yellowstone in the first place, consider this: Soon after Yellowstone was established, a bill was introduced to de-establish it. One of the visionaries, Senator George Graham Vest of Missouri, needed and used all his clout to defeat it. And then it was re-introduced. Sen. Vest did it again. And then it was introduced a third time. Sen. Vest must have sighed, but he did not give up. He beat it back once more.[10]

Are we now to abandon our Yellowstones? Has the concept of the national park been a 100-year wonder that has outlived its usefulness? We know our national parks are beset by problems. Population pressures have whittled away the land that surrounds them and endangered the larger animals that must cross reserve boundaries to stay alive. Too many tourists nose about in the secluded sancta. Climate changes threaten to pull the weather rug out from under the reserves' precious habitats. And remember, in chapters 8 and 9 the science of biodiversity showed us that our national parks are too small anyhow. **O woe! Woe and despair!**

Never mind. Even in a world of reconciliation ecology, we are going to need our Yellowstones. A biotic reserve system anchored by jewels such as Yellowstone will still be able to contribute immensely to the conservation of diversity. But it must give up certain fictions to do so.

We humans now dominate massive fractions of the Earth's surface. The biotic reserve system of the new millennium must admit that and work within the ominous constraints of that domination. It must carve out a new role for reserves in the face of potentially destructive climate changes. It must continue to develop ways to manage its lands, its waters and their wild populations, ignoring completely the misguided hue and cry of those who claim that humans never did any good for any landscape. It must, in short, give up the romantic fantasy that somehow our reserves can function to preserve antique nature in miniature. No one would be happier than I if they could. But the science of species-area relationships proves that they cannot. We will destroy their usefulness if we fly in the face of scientific reality and insist that they can.

However, before we consider what they can do in their second century, let's give them the credit they have earned for what they did in their first. Imagine that only a single 5 percent chunk of the contiguous USA remained unplundered. No matter where we put it, it will miss most of the nation's habitats. A 5 percent area is about the size of the state of Montana. And it is only a hair smaller than the state of California. Yes, both of those states have majestic vistas and many species. But imagine the biotic disaster if, out of the entire lower 48 states, only one of those two states remained. I have yet to hear of a breeding population of redwoods in Montana, or one of balsam poplars in California, or one of any sort of magnolia in either state. Where are their prothonotary warblers, their banner-tailed kangaroo rats, their pine voles, their Diana fritillaries, Palamedes swallowtails, and silvery blue butterflies? No doubt about it, we'd have quite a biotic disaster indeed.

Now consider what we actually do have: a semi-carefully scattered set of reserves, sometimes strategically located to protect a rich, rare, and beautiful set of habitats. We have an Everglades National Park, a Saguaro National Park, a Great Smokies National Park, a Sequoia, a Pipestone, a Cape Cod. And of course, we have Yellowstone itself! What we have is much better than a random single chunk.

How much better have we done? Except for freshwater species and those unfortunate terrestrial species that disappeared before we even got to know them, we have managed to embrace almost the entire country's diversity in our reserve system. We have likely put off the extinctions of 10 percent or even 20 percent of our treasure.

But "put off" is the operative verb. We know that some of the species left after the bulldozer's blade has passed have no future—even some of those that are now alive in our reserves. True, they have a few populations left, but all of these live in habitat that is less than adequate to support them.

Recall that population ecologists term such habitats "sinks" and that I call the species living in them "sink species." The opposite are "source species." Sink species have diseased demographics. Their average rate of generation-to-generation replacement is less than one. They are extinctions-in-waiting, their species, zombies—the living dead. They populate our *Red Book* of the threatened and endangered.[11]

Reconciliation ecology has the job of restoring species demographics to health. But each act of reconciliation produces a habitat different from any natural habitat. None of them is a wilderness. Wearing ballet slippers when we walk the land is not the same as going barefoot. The careful foot is not bare. That reality makes each reserve even more precious. Only our reserves will preserve a semblance of the ancient habitats in which we and our fellow Earth travelers grew up.

Yet, in a world of reconciliation ecology, the role of reserves must change dramatically because we need them as much more than ornaments. We need them to do their share of the work. The trouble is that many species, no matter how much we try to help them, will not be able to persist in our novel, reconciled habitats.

We already noticed such species in South Africa's apple-growing district. Six bird species, such as the orangebreasted sunbird, survived only in the heathland reserve. We also saw it in the thrush family. Many thrushes, like Townsend's solitaire, may require the most natural habitats that we can protect. Such species, the world's unwavering, resolute and steadfast *kulturmeiders*, will survive in our reserves or not at all. This pressure will lead us to alter drastically the ways in which we manage our reserves. It will even change our whole philosophy of the ideal reserve.

We can manage our reserves to maximize the number of species they save by thinking about the species that will most need them. Species that must live in reserves, shoehorned into the tiny leftovers of nature that human society sets aside for them, will probably have small populations and stand a higher risk of extinction from accidents. Often these "wilderness" species are already rare and local and spectacular. On the other hand, species that reconciled habitats can help will often have very large populations because they will live all over the large areas in which we live.

But the common species will not care about the plight of their relatives. Although abundant in reconciled habitats, many will also thrive in reserve habitats and be abundant in them. Even if a reserve offers them marginal habitat, common species may enter its edges as sink populations. In either case, the common species will be wasting precious resources that could have helped the rarer ones by increasing their populations. Although it may be very difficult to stop in practice, we should try to prevent the common species from doing that. In a reconciled world, we may need to rid the reserves of some species that are common outside them.

It sounds like another ecological curve ball, and it is. To save the most species, we may need to eliminate some from our reserves. A radical notion? Yes. But in difficult times we may need to consider radical answers. Yellowstone's very name stands for foresight and innovation in conservation biology. It is a paradox, but what better way to maintain her tradition than to ask how we need to manage Yellowstone differently?

Today, surrounded and inundated by a rising tide of people and the works of people, our Yellowstones find themselves inadequate and menaced. But, unlike King Canute, they cannot command the tide to roll back. They must instead seek higher ground. It won't be easy, but if we work together, the heights are there for us to claim.

Reconciliation Ecology Works in Rich and Poor Economies

When I speak to people about reconciliation ecology, I invariably hear the complaint that conservation is for rich folks only. Luckily, that is not true of reconciliation ecology. Many of the world's conservation schemes collide with the stone wall of economic reality. But reconciliation ecology does not. It joins with the works of humans. It enhances them and may even make them more sustainable and profitable in the long run.

Reconciliation ecology will spawn industries to serve it and to make it even more lucrative. Researchers will design the habitats; landscapers will create them. Seed suppliers and nurseries will supply the rare plants, and also supply ready-to-plant native species to hurry along the new habitats on their path to maturity. Manufacturers and suppliers of an infinity of materials will make it easier to get what we want. Gardeners will maintain our new habitats. Population biologists will monitor the enterprise. Controlling our habitats has always been a big business and it will continue to be so when its goal is reconciliation.

Nevertheless, don't let all this big business talk lead you to believe that reconciliation ecology ignores most of the world's places and peoples. True, those landscape architects better stick to the world's rich countries. The poor ones cannot afford them yet. But reconciliation ecology is just as relevant and applicable in poorer societies as in richer ones.

You see, a surprising lesson emerges from examples of reconciliation. Traditional ways of exploiting the Earth, often associated with having less cash, tend to be sustainable and harbor many native species. Shade coffee plantations, family farms and ranches, small timber holdings, vicuña shearing. Mahdav Gadgil[12] goes so far as to claim that where people have lived in a place for a long time and have been in complete control of their own resource base, they tend also to adopt practices that promote both sustainable use of those resources and conservation of diversity. Gretchen Daily has found so many cases of this pattern that she has given it a name: countryside biogeography.[13]

Often diversity gets into trouble when such traditional modes of exploitation get replaced by high-tech, highly mechanized, big business ones. Such methods rarely consider the question of sustainability. They want profit, as much as possible and calculated over short periods like years or even quarters. This is slash-and-burn exploitation.

What we want, what we must have, is a world that can feed and sustain us over the course of generations. Most likely we have God to thank that such a world will also sustain the largest possible species diversities.

Reconciliation Ecology and the Shifting Baseline Syndrome

Now we come to the secret weapon of reconciliation, its two-edged sword: Human beings can get used to almost anything. And what we get used to, we come to prefer.

My students were the first to teach me that. The subject was bread.

It was 1967 and we were working with our pocket mice and kangaroo rats in lovely Ramsay Canyon, five miles from the town of Sierra Vista, Arizona. In those days, Sierra Vista was tiny, far too small to support its own bakery. We had to use—"use" seems more accurate than "eat"— white, balloon bread. Well, its label said it was nutritious.

On Fridays, a shipment of real bread came in from Tucson, some 90 miles away. By the time it arrived, it had passed well beyond oven-freshness, but my wife Carole made a special trip to fetch it back to the canyon for lunch anyway. She set it out as a surprise, but the students were not pleased. They did try it, but admitted that they greatly preferred the usual air-loaf.

I could not believe it. I guessed there had to be something wrong with their genes. Or maybe they were suffering from lead poisoning? But they merely wanted what they always had. And if you think there is anything extraordinary about them, check out the hordes of American tourists in Paris, crowding into MacDonald's or the Burger King on the Champs-Élysées. In Paris, of all places!*

Environments affect us the same way. The environments we have gotten used to over the past few centuries have deeply eroded our horizons. We have adapted to concrete and we expect noxious fumes. We are comforted by sterility, and we anticipate *not* seeing much around us except for sparrows and a few house plants. When sometimes we do "get away from it all," by taking a trip to a national park or reserve, most of us depend on its wild things being in predictable places at predictable times. In our monster four-wheel-drive sport-utility vehicles, we carry our luxuries with us—our synthetic-fiber camping gear, our thousand-dollar mountain bikes, our boom boxes, our cell phones. We do not seek truly to immerse ourselves in the world of Nature, but rather to use it as an exotic backdrop for artificial diversions. We disport ourselves in high-speed pleasure boats with deafening motors. We race around aimlessly in all-terrain vehicles. And when, thoroughly exhausted, we do allow the din to settle down, we fill the silence with the sounds of our own trivial chatter. We have disconnected ourselves from the world of Nature and have learned to prefer it that way. Nature makes us uneasy, even fearful.

But primarily, Nature no longer holds promise for us. Not of abundance. None of diverse, fertile pulsation. None of sustenance. Having conquered Nature, we have lost both our esteem for her and our faith in her wealth. She has become a meager shadow of herself, and we are aware of it. We have stopped believing in her robustness because we can no longer remember it. We are like great-grandchildren who cannot imagine that their toothless, weak, withered old great-grandmother ever had sex, carried children, or flowed with strength.

Fisheries biologist Daniel Pauly gave a name to this failure of intergenerational memory. He called it the "shifting baseline" syndrome.[14]

---

*The French call the stuff *malbouffe* (garbage food). They eat it by the ton. But at least, what they call it tells us that they know what they are doing.

I tried to capture a sense of the shifting baseline in the memoir following chapter 5. There, I also tried to crystallize it in an epigraph: "It's easy for people not to miss what they never had and never saw." Thus put, who could disagree? But the shifting baseline syndrome is far from trivial and, like many great scientific insights, it took genius to notice.

Pauly became aware of the shifting baseline syndrome after hearing hundreds of fish stories. Each described large catches of now rare or totally extirpated fishes. One story absolutely beggars the imagination.

The grandfather of Pauly's coworker was a fisherman in the 1920s. He drew his catch of mackerel from the waters of the Kattegat, an arm of the sea between Denmark and Sweden. It must have been a hard life because he was plagued by numerous bluefin tuna that entangled themselves in his nets. At the time, no one cared much about bluefin tuna; they were just a common pest. Today, of course, they are rarely, if ever, seen in the North Sea. In those few places in the world's oceans where their dwindled schools do remain, experts meticulously monitor their population biology, and nations carefully apportion the right to fish them. Called *honmaguro* in Japanese, they are a delight of sushi eaters. In early January 2001, a single 200 kilogram bluefin tuna—just one fish—sold at the Tsukiji Central Fish Market auction for some $175,000 (U.S.). That amounts to $875 per kilo, the record price (so far), but good bluefin does typically sell for $200 and up per kilo.[15]

A few more tales of the past will help you understand what we have lost. Jeremy Jackson found evidence of our shifted baseline in the Caribbean.[16] Today, there and everywhere, sea turtles of all species are rare or threatened. But a few hundred years ago, green turtles were so abundant that ships struck vast shoals of them—and sank! And manatees aplenty, along with teeming multitudes of man-sized herbivorous fishes, kept the sea grasses closely cropped. Today they are thick and tall.

Here's another story. An edible mussel, called the giant floater, lives in the bottoms of some North American freshwater streams and lakes. It is called "giant" for the simple reason that it quickly grows to be about a foot long. Archaeologists find great heaps of its shells in the middens of the first Americans. Not too many generations ago, it was so abundant that it was a staple food in many places in the middle of the continent. Brandauer and Wu estimate its population densities at six to twelve per square foot.[17] You can guess that people just collected dinner while they were bathing.

Yes, Virginia, this is Brooklyn! At least it was Brooklyn about 1925 when Abe Cohen took this photo near his home. Unfortunately, Mr. Cohen has passed away. Along with him and his generation goes any possibility that anyone can imagine Brooklyn, New York, ever looking so sweet and green. (Thanks to Mel Cohen and Molly Senor for permission to scan the photo.)

Today's populations of giant floaters, like the majority of North American freshwater animals, are going, going, and gone. In the waters of Colorado, they exist at population densities of less than one per hundred square feet in the few sites where they still survive at all.[18] Thus, they are more than a thousand times scarcer than they were only a century or two ago.

Finally, consider the extinction of North America's principal pigeon, the passenger pigeon. Two hundred years ago, it numbered in the billions. Audubon wrote of a single pigeon multitude that, near Cincinnati in 1813, took three days to fly past him, all the while so dense that it darkened the heavens.[19] In Audubon's words: "The light of the noonday was obscured as by an eclipse." And those birds flew fast—a clever experiment of his showed that they flew about a mile a minute. There were more than a billion birds in that flock.

In 1800, approximately every other bird in the continent was a passenger pigeon. And oh, we did eat them! Professional pigeoners shot them in hordes and supplied the cities of the eastern seaboard with fresh pigeon meat for a century.

We lived in a culture of pigeons, so much so that the vocabulary entered our speech where it remains to this day. A "stool pigeon" was one tied to a stool in a field so that its cries to fly free would attract other pigeons into the sights of the pigeoner's rifle. A "clay pigeon" is a substitute target for the real thing. You might say that we took pigeons and their abundance for granted.

But they are gone and their world is gone and we can never imagine what it was like. That is the point. Our ancestors lived on an Earth where they took Nature's abundance and diversity for granted. We live on one where we take her fragility and poverty for granted. We simply have not experienced enough to know how different she could be.

Novelist John McDonald was no ecologist, but he saw the shifting baseline syndrome all around him in southern Florida. He wrote:

> Florida can never really come to grips with saving the environment because a very large percentage of the population at any given time just got here. So why should they fight to turn the clock back? It looks great to them the way it is. Two years later, as they are beginning to feel uneasy, a few thousand more people are just discovering it all for the first time and wouldn't change a thing. And meanwhile the people who knew what it was like twenty years ago are an ever-dwindling minority, a voice too faint to be heard.[20]

And, indeed, our world does not look so poor to us when we are newly arrived in it. Too bad we cannot bequeath our memories to our children. Too bad we cannot find some way for them to see the forests and meadows that we saw when we were younger. If we could, they would be outraged at what they have lost. And they would resolve to recover it somehow. When we die, our children inherit not only our worldly possessions, but the world's environments—in whatever deteriorated condition we allow them to emerge from our watch. Sadly, we must take our memories of a richer world along with us. What a tragedy.

Yet the shifting baseline syndrome works both ways. True, degrading our environment causes us to expect less of it. But improving our environment will cause us to expect more. In a world full of life, we will come to rejoice in life. Surrounding ourselves with our cousins from other species, we will no longer acquiesce in sterility. In neighborhoods full of diversity, we will demand even more diversity and work to get it.

I suppose I see reconciliation ecology as the great environmental educator. The environments it creates will teach us once again how to take pleasure from Nature. The species we live with will resensitize us to her delights and addict us to her bounty. They will spur us on. We will wonder how we ever could have settled for less, or fooled ourselves into believing that "less" could satisfy us and nurture us and give us fair value.

As the new millennium opened, the media engorged on visions of the future. All of them involved technology. Travel to the stars, magical computers, DNA transplants, and grow-your-own replacement organs. I'm not saying we should not want these things. I'm not predicting that we will not get them. But I will venture the guess that none of them will make us much happier.

My own vision of the year 2100 is fundamentally different. We will have the technology but we will be using it in magnificent surroundings, in environments that help us to discharge our responsibilities to nature and bring us peace and fulfillment. I see my great-great-grandchildren opening their doors on a world of wonders. I smell the perfumes of their native shrubs. I hear the chirp of now-scarce warblers in their trees. I see now-vanishing butterflies enchanting their gardens as they sip from once-rare flowers.

A Chinese proverb claims that a cheerful heart makes its own song. Maybe my vision of 2100 is my song, a mere fantasy, my way to cope with the unthinkable. But surely it need not be so. Another Chinese proverb holds, "Enough shovels of earth . . . a mountain. Enough pails of water . . . a river." Similarly, reconciliation ecology thrives incrementally. It does not require the grand design, the universally accepted paradigm. Every little contribution can add meaningfully to the dream. And if enough of us share that dream, we, our children and the millions of species that travel together with us on Planet Earth can live in it.

Notes

Chapter 1

1. *Isaiah* 11:6.
2. Quoted by Merle Miller, 1973. *Plain Speaking; an Oral Biography of Harry S. Truman* (p. 240). Berkeley Publishing/G.P. Putnam's Sons, New York.
3. David Western, 2001. Human-modified ecosystems and future evolution. *Proceedings of the National Academy of Science (USA)* 98: 5458–5465.
4. Lewis Mumford, 1940. *Survey Graphic.* Quoted in Wes Jackson, 1985. *New Roots for Agriculture*, New edition (p. 48). University of Nebraska Press, Lincoln.

Chapter 2

1. Judith H. Heerwagen and Gordon H. Orians, 1993. Humans, habitats, and aesthetics. Ch. 4 (p. 140) in *The Biophilia Hypothesis*, Stephen R. Kellert and Edward O. Wilson (eds.). Island Press, Washington, DC.
2. M. L. Rosenzweig, 1973. Habitat selection experiments with a pair of coexisting heteromyid rodent species. *Ecology* 54: 111–117.
3. W. Whyte, 1988. *City: Rediscovering the Center* (p. 123). Doubleday, New York.
4. Barbara Ward and René Jules Dubos, 1972. *Only One Earth: The Care and Maintenance of a Small Planet* (p. 101). United Nations Conference on the Human Environment (1972: Stockholm, Sweden). W. W. Norton, New York.
5. Judith H. Heerwagen and Gordon H. Orians, 1993. Humans, habitats, and aesthetics. Ch. 4 in *The Biophilia Hypothesis*, Stephen R. Kellert and Edward O. Wilson (eds.). Island Press, Washington, DC; Gordon H. Orians, 1998. Human behavioral ecology: 140 years without Darwin is too long. *Bulletin of the Ecological Society of America* 79(1): 15–28.
6. Craig Tufts and Peter Loewer, 1995. *Gardening for Wildlife*. Rodale Press, Emmaus, PA.
7. Herbert Bormann, Diana Balmori, and Gordon Geballe, 1993. *Redesigning the American Lawn.* Yale University Press, New Haven, CT.
8. Ibid., p. 138.
9. www.nwf.org
10. Joseph Mailliard, 1930. *Handbook of the Birds of Golden Gate Park San Francisco.* California Academy of Sciences, San Francisco.

11. Carla Cicero, 1989. Avian community structure in a large urban park: controls of local richness and diversity. *Landscape and Urban Planning* 17: 221–240.

12. Manfred Köhler, 1990. The living conditions of plants on the roofs of buildings. Pp. 195–207 in *Urban Ecology*, Herbert Sukopp and Slavomil Hejny (eds.). International Botanical Congress (14th: 1987: Berlin, Germany). SPB Academic Publishing, The Hague.

Chapter 3

1. *Numbers* 31:23.

2. Brenda Biondo, 1997. In defense of the longleaf pine. *Nature Conservancy* 47 (4; September/October): 10–17.

3. U.S.D.A. Forest Service, 1997. *Report of the United States on the Criteria and Indicators for the Sustainable Management of Temperate and Boreal Forests* (6 June 1997, pp. 2–3, 2–12). Washington, DC.

4. R. W. McWhite, D. R. Green, C. J. Petrick, S. M. Seiber, and J. L. Hardesty, 1993. *Natural Resources Management Plan, Eglin Air Force Base, Florida.* U.S. Dept. of the Air Force, Eglin Air Force Base, FL.

5. Emily Dickinson, 1884. Poem 1641. *The Poems of Emily Dickinson; Reading Edition,* R. W. Franklin (ed.). Belknap Press of Harvard University Press, Cambridge, MA.

6. R. W. McFarlane, 1992. *A Stillness in the Pines. The Ecology of the Red-cockaded Woodpecker.* W. W. Norton, New York.

7. Biondo, see note 2.

8. David S. Wilcove, Michael J. Bean, Robert Bonnie, and Margaret McMillan, 1996. *Rebuilding the Ark; Toward a More Effective Endangered Species Act for Private Land.* Environmental Defense Fund, New York.

Chapter 4

1. Quoted in George Seldes, 1967. *The Great Quotations* (p. 667). Pocket Books, New York.

2. I owe the latter two sentences to Robert Gordis (1944, Jewish Law and Catholic Israel, *Proc. Rabbinical Assembly* 41–44; 64–93). I confess to taking them entirely out of context.

3. Gretchen C. Daily (ed.), 1997. *Nature's Services: Societal Dependence on Natural Ecosystems.* Island Press, Washington, DC.

4. Genesis 6: 19.

5. Genesis 8: 17.

6. George Perkins Marsh, 1855. *The Camel: His Organization, Habits and Uses* (pp. 98–122). Ninth Annual Report of the Smithsonian Institution for 1854. U.S. Senate, Washington, DC. And Marsh, 1860. The study of nature. *Christian Examiner* 68: 33–62. Both may be found in *So Great a Vision; the Conservation Writings of George Perkins Marsh,* 1999, Stephen C. Trombulak (ed.). Middlebury College Press, Hanover, NH.

7. Most notably: Lynn White, Jr., 1967. The historical roots of our ecological crisis. *Science* 155: 1204–7; Ian L. McHarg, 1969. *Design with Nature.* Natural History Press, Garden City, New York.

8. George Perkins Marsh, 1864 [1965]. *Man and Nature; or, Physical Geography as Modified by Human Action.* David Lowenthal (ed.). Belknap Press of Harvard University Press, Cambridge, MA.

9. V. M. Edwards, 1995. *Dealing in Diversity: America's Market for Nature Conservation.* Cambridge University Press, Cambridge, U.K.

10. Eric Hoffman, 1999. Managing the vicuna and its golden fleece. *The Alpaca Registry Journal* 4(1): http://www.alpacaregistry.net/journal/spr99j_07.html.

11. J. Rabinovics, A. Capurro, and L. Pessina, 1991. Vicuña use and the bioeconomics of an Andean peasant community in Catamarca, Argentina. Pp. 337–358 in *Neotropical Wildlife Use and Conservation,* J. Robinson and K. Redford (eds.). University of Chicago Press, Chicago.

12. C. Bonacic, D. W. MacDonald, J. Galaz, and R. M. Sibly, 2002. Density dependence in the camelid *Vicugna vicugna*: the recovery of a protected population in Chile. *Oryx* 36: 118–125.

13. Maud and Miska Petersham, 1948. *The Story Book of Wool.* John C. Winston, Philadelphia.

14. Mark Cherrington, 1999. The falcon and the firebrand. *Earthwatch* May/June, 39–47.

15. Paul Faeth, R. Repetto, K. Kroll, Q. Dai, and G. Helmers, 1991. *Paying the Farm Bill: U.S. Agricultural Policy and the Transition to Sustainable Agriculture.* World Resources Institute, Washington, DC.

16. R. L. Doutt and J. Nataka, 1973. The *Rubus* leafhopper and its egg parasitoid: an endemic biotic system useful in grape pest management. *Environmental Entomology* 2: 381–386.

17. K. M. Kuruvilla, V. V. Radhakrishnan, and K. J. Madhusoodanan, 1995. Small cardamom plantations—floristic calendar and bee pasturage trees. *Journal of Environmental Resources* 3: 32–33. Thanks to Margie Mayfield for this reference.

18. N. Dover and L. Talbot, 1987. *To Feed the Earth: Agroecology for Sustainable Development.* World Resources Institute, Washington, DC.

19. *Arizona Daily Star,* Tucson, AZ, 17 Feb. 2000.

Chapter 5

1. Julian Barnes, 1989. *A History of the World in 10½ Chapters* (p. 29). Jonathan Cape, London.

2. Wes Jackson, 1980. *New Roots for Agriculture* (p. 18). Friends of the Earth, San Francisco.

3. Paul Faeth, R. Repetto, K. Kroll, Q. Dai, and G. Helmers, 1991. *Paying the Farm Bill: U.S. Agricultural Policy and the Transition to Sustainable Agriculture.* World Resources Institute, Washington, DC.

4. Norman Myers and J. Kent, 1998. *Perverse Subsidies.* Institute for Sustainable Development, Winnipeg, Saskatchewan.

5. E. A. Norse, 1990. *Ancient Forests of the Pacific Northwest.* Island Press & The Wilderness Society, Washington, DC.

6. John Vandermeer and Ivette Perfecto, 1995. *Breakfast of Biodiversity: The Truth about Rain Forest Destruction.* Institute for Food and Development Policy, Oakland, CA; Robert A. Rice and Justin R. Ward, 1996. *Coffee, Conservation, and Commerce in the*

Western Hemisphere. Natural Resources Defense Council & Smithsonian Migratory Bird Center, Washington, DC.

7. Ibid., 144.

Chapter 6

1. Quoted in George Seldes, 1967. *The Great Quotations* (p. 782). Pocket Books, New York.

2. R. M. Little and T. M. Crowe, 1994. Conservation implications of deciduous fruit farming on birds in the Elgin district, Western Cape Province, South Africa. *Transactions of the Royal Society of South Africa* 49: 185–198.

3. Thanks to Harald Beck for calling the words to my attention.

4. E. Howard Eaton, 1914. *Birds of New York* (vol. 2) *Memoirs of the New York State Museum* (pp. 258, 259). University of the State of New York, Albany.

5. Lawrence Zeleny, 1976. *The Bluebird: How You Can Help Its Fight for Survival.* Indiana University Press, Bloomington.

6. A good source: Wayne H. Davis and Phillippe Roca, 1995. *Bluebirds and Their Survival.* University Press of Kentucky, Lexington.

7. nabluebirdsociety.org

8. Reuven Yosef and F. E. Lohrer (eds.), 1995. *Shrikes (Laniidae) of the World: Biology and Conservation. Proceedings of the Western Foundation of Vertebrate Zoology* 6:1–343.

9. Reuven Yosef and T. C. Grubb, Jr., 1994. Resource dependence and territory size in loggerhead shrikes (*Lanius ludovicianus*). *Auk* 111: 465–469.

10. Conservation biologists still speak of *The Red Book of Rare and Endangered Species* although this title is no longer strictly accurate. Today all countries keep a separate red book that applies to their own threatened species, and the central list has been transferred to the internet (The IUCN Red List of Threatened Species, www.redlist.org). But the title as well as the term "red book species" is current in conservation discussions and publications.

11. Claire Devereux, 1998. The fiscal shrike. *Africa—Birds & Birding* 3: 52–57.

12. Dries van Nieuwenhuyse, 1998. Conservation opportunities for the red-backed shrike (*Lanius collurio*). Pp. 79–82 in *Shrikes of the World – II: Conservation Implementation,* Reuven Yosef and F. Lohrer (eds.). International Birdwatching Center in Eilat, Eilat, Israel.

13. Martin Schön, 1998. Conservation measures and implementation for the great grey shrike (*Lanius excubitor*) in the southwestern Schwäbische Alb of southwestern Germany. Pp. 79–82 in *Shrikes of the World – II: Conservation Implementation,* Reuven Yosef and F. Lohrer (eds.). International Birdwatching Center in Eilat, Eilat, Israel.

14. J. S. Denton, S. P. Hitchings, T. J. C. Beebee, and A. Gent, 1997. A recovery program for the natterjack toad (*Bufo calamita*) in Britain. *Conservation Biology* 11: 1329–1338.

15. The work of Phillip R. Rosen and Cecil Schwalbe appears in three detailed but unpublished reports (1996) to Arizona Game & Fish Department (Heritage Program) and U.S. Fish & Wildlife Service.

16. Bill McDonald, 1996. A working wilderness. *Endangered Species Bulletin* 21(1): 22–23.

17. Mitch Tobin, 14 June 2002. Leopard frog gets protection. *Arizona Daily Star,* B4–B5.

18. Verlyn Klinkenborg, 1995. Crossing borders. *Audubon* September–October: 34–47; Mark Cheater, 1995. Good guys in the badlands. *Nature Conservancy* July–August: 16–23.

Chapter 7

1. Jean Baptiste Poquelin, 1670. *Le Bourgeois Gentilhomme*, Act II, Scene 4.
2. *Proverbs* 10: 22.
3. *Deuteronomy Rabbah* 3: 3.
4. R. Gaby, M. P. McMahon, F. J. Mazzoti, W. N. Gillies, and J. R. Wilcox, 1985. Ecology of a population of *Crocodylus acutus* at a power plant site in Florida. *Journal of Herpetology* 19: 189–198.
5. H. Esselink, M. Geertsma, L. Kuper, F. Hustings, and H. van Berkel, 1995. Can peat-moor regeneration rescue the red-backed shrike in the Netherlands? *Shrikes (Laniidae) of the World: Biology and Conservation Proceedings of the Western Foundation of Vertebrate Zoology* 6: 287–293.
6. Robert Dulfer and Kevin Roche (eds.), 1998. *First Phase Report of the Třeboň Otter Project. Nature & Environment*, No. 93. Council of Europe Publishing, Strasbourg, France. Except for one other source, the entire Třeboň section is drawn from this publication.
7. J. Janda, 1994. Třeboň Basin Biosphere Reserve. Pp. 65–80 in *Biosphere Reserves on the Crossroads of Central Europe*, J. Jeník and M. F. Price (eds.). Czech National Committee for UNESCO's Man and Biosphere Programme, Prague.
8. Contact Bat Conservation International, www.batcon.org
9. Russell Greenberg, P. Bichier, and J. Sterling, 1997. Acacia, cattle and migratory birds in southeastern Mexico. *Biological Conservation* 80: 235–247.

Chapter 8

1. *Locksley Hall Sixty Years After* (1886)
2. F. H. A. von Humboldt, 1807. *Essai sur la geographie des plantes*. Von Humboldt, Paris.
3. Ibid.
4. Quoted in J. G. Dony, 1963. The expectation of plant records from prescribed areas. *Watsonia* 5: 377–385.
5. M. L. Rosenzweig, 1995. *Species Diversity in Space and Time*. Cambridge University Press, Cambridge, UK.
6. Irene C. Wisheu, Michael L. Rosenzweig, Linda Olsvig-Whittaker, and Avi Shmida, 2000. What makes nutrient-poor Mediterranean heathlands so rich in plant diversity? *Evolutionary Ecology Research* 2: 935–955.
7. C. B. Williams, 1943. Area and the number of species. *Nature* 152: 264–267.
8. Wade A. Leitner and M. L. Rosenzweig, 1997. Nested species-area curves and stochastic sampling: a new theory. *Oikos* 79: 503–512.
9. Brian J. McGill, 2002. A spatial mechanism for macroecological patterns (abstract). Annual meeting of the Ecological Society of America, Tucson, AZ. URL: 199.245.200.45/pweb/document/?SOCIETY=esa&YEAR=2002&ID=5820.
10. Robert H. MacArthur and Edward O. Wilson, 1967. *The Theory of Island Biogeography*. Princeton University Press, Princeton, NJ.
11. Edward O. Wilson, 1961. The nature of the taxon cycle in the Melanesian ant fauna. *American Naturalist* 95: 169–193; Edward O. Wilson and R. W. Taylor, 1967. *The ants of Polynesia (Hymenoptera: Formicidae). Pacific insects monographs*, No. 14. Entomology Dept., Bernice P. Bishop Museum, Honolulu, HI.

Chapter 9

1. Quoted in John Bartlett, 1992. *Familiar Quotations*, 16th edition (p. 741, note 2). Justin Kaplan (ed.). Little, Brown, Boston, MA.

2. Jeremy B. C. Jackson, 2001. What was natural in the coastal oceans? *Proceedings of the National Academy of Science (USA)* 98: 5411–5418.

3. U.S. Department of Interior and U.S. Department of Agriculture, 1992. *America's Biodiversity Strategy: Actions to Conserve Species and Habitats*. Washington, DC.

4. R. F. Noss and A. Y. Cooperrider, 1994. *Saving Nature's Legacy*. Island Press and Defenders of Wildlife, Washington, DC.

5. Ibid., p. 253.

6. Ibid., p. 172.

7. Ibid., p. 195.

8. U.S.D.A. Forest Service, 1997. *Report of the United States on the Criteria and Indicators for the Sustainable Management of Temperate and Boreal Forests* (6 June 1997, p. 1–4). Washington, DC.

9. *Atlas de evolução dos remanescentes florestais e ecossistemas associados no dommnio do Mata Atlântica no permodo*. Fundação SOS Mata Atlântica, Instituto Nacional de Pesquisas Espaciais & Instituto Socioambiental, Saõ Paulo, SP, Brazil. Thanks to Gustavo da Fonseca for bringing this reference to my attention and telling me what it says.

10. William Wayt Thomas and André Maurício V. de Carvalho. 1993. Estudo fitossociologico de Serra Grande, Uruçuca, Bahia, Brasil. *XLIV Congresso Nacional de Botânica, São Luis, 24–30 de Janeiro de 1993, Resumos* 1: 224. Sociedade Botânica do Brasil, Universidade Federal de Maranhão.

11. José Rezende Mendona, André Maurício V. de Carvalho, Luiz Alberto Mattos Silva, and William Wayt Thomas. 1994. *45 Years of Land Clearing in Southern Bahia*. Herbário Centro de Pesquisas do Cacau, Ilhéus, BA, Brazil & The New York Botanic Garden, New York.

12. J. T. Carlton, G. J. Vermeij, D. R. Lindberg, D. A. Carlton, and E. C. Dudley, 1991. The first historical extinction of a marine invertebrate in an ocean basin: the demise of the eelgrass limpet *Lottia alveus*. *Biological Bulletin* 180: 72–80.

13. R. N. Holdaway, 1989. New Zealand's pre-human avifauna and its vulnerability. *New Zealand Journal of Ecology* 9: 11–25.

14. D. J. Kitchener, A. Chapman, J. Dell, B. G. Muir, and M. Palmer, 1980. Lizard assemblage and reserve size and structure in the Western Australian wheatbelt—some implications for conservation. *Biological Conservation* 17: 25–62.

15. D. J. Kitchener, J. Dell, B. G. Muir, and M. Palmer, 1982. Birds in Western Australian wheatbelt reserves—implications for conservation. *Biological Conservation* 22: 127–163.

16. Bruce A. Wilcox, 1978. Supersaturated island faunas: A species-age relationship for lizards on post-pleistocene land-bridge islands. *Science* 199: 996–998.

17. M. L. Rosenzweig, 1998. Preston's ergodic conjecture: the accumulation of species in space and time. Ch. 14 in *Biodiversity Dynamics: Turnover of Populations, Taxa and Communities*, M. L. McKinney and J. Drake (eds.). Columbia University Press, New York.

Chapter 10

1. Douglas Adams, 1984. *So Long and Thanks for All the Fish* (p. 121). Harmony Books, New York.

2. R. L. Peters and J. D. S. Darling, 1985. The greenhouse effect and nature reserves. *Bioscience* 35: 707–717.

3. W. William Weeks, 1997. *Beyond the Ark; Tools for an Ecosystem Approach to Conservation*. Island Press, Washington, DC.

4. David S. Wilcove and Linus Y. Chen, 1998. Management costs for endangered species. *Conservation Biology* 12: 1405–1407.

5. Red Book, see chapter 6, note 10.

6. Ambrose Bierce, 1898. The flying-machine. In *Fantastic Fables*. G. P. Putnam's Sons, New York.

7. National Research Council, 1995. *Science and the Endangered Species Act*. National Academy Press, Washington, DC.

8. James C. Greenway, Jr. 1958. *Extinct and Vanishing Birds of the World*. Special Publication 13, American Committee for International Wildlife Protection, New York.

9. Data courtesy of David Wilcove from David S. Wilcove, Michael J. Bean, Robert Bonnie, and Margaret McMillan, 1996. *Rebuilding the Ark; Toward a More Effective Endangered Species Act for Private Land*. Environmental Defense Fund, New York.

10. John Vandermeer and Ivette Perfecto, 1995. *Breakfast of Biodiversity: The Truth about Rain Forest Destruction* (p. 140). Institute for Food and Development Policy, Oakland, CA.

Chapter 11

1. Joseph Wood Krutch, 1951. *The Desert Year* (p. 214). William Sloane Associates, New York.

2. Quoted in Robert West Howard, 1975. *The Dawn Seekers: The First History of American Paleontology* (p. 56). Harcourt Brace Jovanovich, New York.

3. *Arizona Daily Star*, 24 September 2001.

4. David Steadman, 1995. Prehistoric extinctions of Pacific Island birds: biodiversity meets zooarchaeology. *Science* 267: 1123–1131.

5. Richard N. Holdaway and Christopher Jacomb, 2000. Rapid extinction of the moas (Aves: Dinornithiformes): model, test, and implications. *Science* 287: 2250–2254.

6. David W. Steadman, Patricia Vargas Casanova, and Claudio Cristino Ferrando, 1994. Stratigraphy, chronology, and cultural context of an early faunal assemblage from Easter Island. *Asian Perspectives* 33: 79–96.

7. Georgia Lee, Easter Island Foundation.

8. J. R. Flenley, 1993. The paleoecology of Easter Island, and its ecological disaster. Pp. 27–45 in *Easter Island Studies*, S. R. Fischer (ed.). Oxbow Monographs No. 32, Oxbow Books, Oxford, U.K.

9. Historical details from B. G. Corney (translator and editor), 1908. *The Voyage of Captain don Felipe Gonzalez in the Ship of the Line San Francisco, with the Frigate Santa Rosalia in Company, to Easter Island in 1770–1771. Preceded by an Extract from Mynheer Jacob Roggeveen's Official Log of His Discovery of and Visit to Easter Island in 1722*. The Hakluyt Society, Cambridge, U.K.; J. M. Brown, 1924. *Riddle of the Pacific*. Unwin, London, U.K.

10. Brown, see note 9.

11. S. Englert, 1970. *Island at the Center of the World; New Light on Easter Island.* Scribner's Sons, New York.

12. Steadman, see note 6.

13. Wes Jackson, 1980. *New Roots for Agriculture* (p. 116). Friends of the Earth, San Francisco.

14. George Gaylord Simpson, 1953. *The Major Features of Evolution* (p. 281). Columbia University Press, New York.

15. David Jablonski, 1986. Background and mass extinctions: the alternation of macroevolutionary regimes. *Science* 231: 129–133; Douglas Erwin, 1998. The end and the beginning: recoveries from mass extinctions. *Trends in Ecology & Evolution* 13: 344–349.

16. Luann Becker, Robert J. Poreda, Andrew G. Hunt, Theodore E. Bunch, and Michael Rampino, 2001. Impact event at the Permian-triassic boundary: evidence from extraterrestrial noble gases in fullerenes. *Science* 291: 1530–1533.

17. Dave Barry, 1987. *Bad Habits* (pp. 163–163). Henry Holt, New York.

18. E. O. Wilson, 1984. *Biophilia: The Human Bond with Other Species* (p. 121). Harvard University Press, Cambridge, MA.

Chapter 12

1. 1862. Poem 314. *The Poems of Emily Dickinson; Reading Edition.* R. W. Franklin (ed.). Belknap Press of Harvard University Press, Cambridge, MA.

2. *Treaty on European Union*, Article 3b.

3. Robert A. Jones, 1996. *The Politics and Economics of the European Union: an Introductory Text.* Edward Elgar, Cheltenham, U.K.

4. John McCormick, 1999. *The European Union: Politics and Policies*, 2nd ed. (p. 87). Westview Press, Boulder, CO.

5. James A. Reinartz, 1995. Planting state-listed endangered and threatened plants. *Conservation Biology* 9: 771–781.

6. 1959. *"Poisoning Pigeons in the Park."*

7. Including W. J. Bond, R. M. Cowling, S. I. Higgins, D. C. Le Maitre, D. M. Richardson, B. W. van Wilgen, and others.

8. René Jules Dubos, 1981. *Celebrations of Life* (p. 81). McGraw-Hill, New York.

9. Barbara Ward and René Jules Dubos, 1972. *Only One Earth; the Care and Maintenance of a Small Planet.* United Nations Conference on the Human Environment (1972 : Stockholm, Sweden). W. W. Norton, New York.

10. Thanks to Lee Whitaker, Yellowstone's historian, for teaching me about this Victorian struggle.

11. Red Book, see chapter 6, note 10.

12. Mahdav Gadgil, F. Berkes, and C. Folke, 1993. Indigenous knowledge for biodiversity conservation. *Ambio* 22: 151–156.

13. Gretchen Daily, Paul R. Ehrlich, and G. A. Sánchez-Azofeifa, 2001. Countryside biogeography: use of human-dominated habitats by the avifauna of southern Costa Rica. *Ecological Applications* 11: 1–13.

14. Daniel Pauly, 1995. Anecdotes and the shifting baseline syndrome of fisheries. *Trends in Ecology & Evolution* 10: 430.

15. Associated Press report, *Arizona Daily Star*, Tucson, 6 January 2001.

16. Jeremy B. C. Jackson, 2001. What was natural in the coastal oceans? *Proceedings of the National Academy of Science (USA)* 98: 5411–5418.

17. Nancy Brandauer and Shi-Kuei Wu, 1978. The freshwater mussels (family Unionidae). *Natural History Inventory of Colorado* 2: 41–60.

18. Hsiu-Ping Lin, Jeffry B. Mitton, and Scott J. Herrmann, 1996. Genetic differentiation in and management recommendations for the freshwater mussel, *Pyganodon grandis* (Say, 1829). *American Malacological Bulletin* 13: 117–124.

19. John James Audubon, 1843. *The Birds of America* (vol. 5, p. 27). Reprinted 1967. Dover, New York.

20. John D. MacDonald, 1978. *The Empty Copper Sea* (p. 20). Fawcett Gold Medal Books, New York.

Acknowledgments

The Morris K. Udall Center for Public Policy supported the initial work on this book during the time I was a Fellow in 1997. The U.S.–Israel Binational Science Foundation helped with some travel expenses. Evolutionary Ecology, Ltd. provided access to its office equipment and supplies. Carole Rosenzweig accompanied me on all but one of the journeys connected with this book. Carole also read the earliest draft (and others) and helped spot many weaknesses, and also helped in obtaining illustrations. Jerry Winakur read the second draft and then provided inestimable advice and moral support. Carole, Jerry, Lee McAden Robinson and I met in Texas to work on the title and subtitle of the book, and to identify and highlight its predominant message. Ron Pulliam's smile and confidence led me to the idea of reconciliation ecology in the first place. David Policansky introduced me to Daniel Pauly's shifting baseline syndrome; I profited greatly from our discussions of it. Many colleagues and students around the world gave me leads that resulted in most of the examples in the book. These include Sarah Armstrong, Alona Bachi, Andrew Balmford, Caleb Gordon, Deborah Gur-Arie, Christine Hawkes, Brian Knauss, John Lawton, Jeff Mitton, Han Olff, Stuart Pimm, Paul Richards and Jerry Winakur. One in particular, Reuven Yosef, provided three examples and the hospitality required to see them first hand in Israel and Florida. I mention others in the credits of illustrations and in endnotes. Thanks to them all.

Illustration Sources

Desert pocket mice, in Seth Benson, 1933. *University of California Publications in Zoology*, Number 40 (frontispiece).

Longleaf pine landscape, in Brenda Biondo, 1997. In defense of the longleaf pine. *Nature Conservancy* 47: 4 (September/October) (p. 15).

Red-cockaded woodpecker in Olin Sewall Pettingill, Jr., 1951. *A Guide to Bird Finding East of the Mississippi* (p. 18). Oxford University Press, New York.

Drilling woodpecker holes, in Brenda Biondo, 1997. In defense of the longleaf pine. *Nature Conservancy* 47: 4 (September/October) (p. 16).

Oldest of all Chinese inscriptions, in Leon Wieger, 1927. *Chinese Characters*, 2nd edition (p. 375). Catholic Mission Press, Ho-kien-fu, China.

Eastern bluebird, in Lawrence Zeleny, 1976. *The Bluebird* (frontispiece). Indiana University Press, Bloomington.

Loggerhead shrike, in Josselyn Van Tyne & Andrew J. Berger, 1959. *Ornithology*, (p. 534). John Wiley & Sons, New York.

Bufo calamita, anonymous illustration in J. E. Taylor, 1889. *The Playtime Naturalist* (p. 194). Chatto & Windus, Picadilly, London, UK.

Otter, anonymous cover illustration for William C. Grimm and Ralph Whitebread, 1952. *Mammal Survey of Northeastern Pennsylvania*. Pennsylvania Game Commission, Harrisburg.

Bird species of different tropical Pacific archipelagos, data from G. H. Adler, 1992. Endemism in birds of tropical Pacific islands. *Evolutionary Ecology* 6: 296–306.

The giant lemur of Madagascar, in Björn Kurtén, 1971. *The Age of Mammals* (p. 211). Weidenfeld & Nicolson, London, UK.

A terror crane, in Björn Kurtén, 1971. *The Age of Mammals* (p. 49). Weidenfeld & Nicolson, London, UK.

Index

Note: In the index, authors' names are followed by the page number on which their work is discussed, whether or not their name actually appears on that page.